Nivalda Batista Santos

Resistance and challenges in securing land ownership

Nivalda Batista Santos

Resistance and challenges in securing land ownership

Quilombola community of Mocambo in the municipality of Porto da Folha/SE

ScienciaScripts

Imprint
Any brand names and product names mentioned in this book are subject to trademark, brand or patent protection and are trademarks or registered trademarks of their respective holders. The use of brand names, product names, common names, trade names, product descriptions etc. even without a particular marking in this work is in no way to be construed to mean that such names may be regarded as unrestricted in respect of trademark and brand protection legislation and could thus be used by anyone.

Cover image: www.ingimage.com

This book is a translation from the original published under ISBN 978-3-330-75785-1.

Publisher:
Sciencia Scripts
is a trademark of
Dodo Books Indian Ocean Ltd. and OmniScriptum S.R.L publishing group

120 High Road, East Finchley, London, N2 9ED, United Kingdom
Str. Armeneasca 28/1, office 1, Chisinau MD-2012, Republic of Moldova, Europe
Printed at: see last page
ISBN: 978-620-8-29245-4

SUMMARY

The black communities and Mocambo for their incessant struggle to legitimize the right to access land.

ACKNOWLEDGMENTS

To Prof. Dr. José Levi Furtado Sampaio for his understanding and patience during my guidance and for sharing his knowledge and generosity.

FUNCAP - for financial support for the research.

To all my family for their unconditional support, especially my mother Lisete for her absence and presence in order to achieve my goals.

To the Mocambo quilombola community, especially to Dona Maria (in memoriam), a representative of the struggle for land titling for the community.

INCRA for making the documents available for consultation. Especially Antonio Carlos Fontenele.

To Prof.ª . Dr.ª . Alexandrina Luz Conceiçâo for her participation in the qualifying exam and for her suggestions on my work. Since I was an undergraduate, she encouraged me to do a master's degree, pointing me in the right direction by believing in my ability. She represents knowledge, wisdom and involvement.

To Prof. Dr. Alexandra Maria de Oliveira for her contributions to the qualification work and for her encouragement for believing in the topic discussed.

To the teachers and fellow students of the Geography Master's course at the Federal University of Cearà.

To all those who make up the administrative body of the UFC Postgraduate Geography Center

LEAT - Laboratory of Agrarian and Territorial Studies.

To LEA - Laboratòrio de Estudos Agràrios and to the Research Group State, Capital, Labor and Territorial Re-ordering Policies.

To the Magalhaes de Sousa family for the warm welcome into their home and for the moments of joy and companionship in placing me as an adopted member of the family.

To Rosana for her support and encouragement, as well as for identifying with the difficulties we went through to achieve our goals.

To Elenilda, Margarette, Maryane, Nilma and Vilma, friends from undergrad, but always present.

To Dayse and Shiziele, a friendship strengthened in the GPECT group.

To Jardélia, a friend for a short time, but always present and with a very strong humorous streak.

Thank you all very much.

"Don't accept what is customary as something natural, because the time of bloody disorder, of organized confusion, of conscious arbitrariness, of dehumanized humanity, nothing should seem natural, nothing should seem impossible to change".

Bertold Brecht

SUMMARY

This dissertation deals with the challenges faced by the remaining community of Quilombo Mocambo, in the municipality of Porto da Folha - SE, in their struggle for recognition under Article 68 of the ADCT (Constitutional and Transitional Provisions Act), in order to regulate land ownership. This dissertation is based on the method of historical materialism, which allowed for an analysis of the contradictions that exist in the countryside with the privatization of land. In order to carry out the proposed analysis, it was necessary to retrieve geographical and historical information about the process of land regulation of quilombola territories in the state of Sergipe, promoted by the federal government through INCRA, as well as the implementation of public and private projects in the areas of quilombola remnants. The research carried out made it possible to reflect on the conflict relations faced by the Mocambo Quilombo throughout history, and to recover the memory of the main subjects who have been historically involved in this struggle for the right to land ownership, obtaining and titling for the maintenance and configuration of the territory of the Mocambo Quilombo remnant community. The set of theoretical and methodological procedures led to the conclusion that it is necessary to guarantee real, non-compensatory policies in the areas of health, education and the economy in order to reduce social and economic inequalities and promote the autonomy of the Mocambo Quilombo.

Keywords: land tenure, territory, territoriality, identity, memory, quilombo

1 INTRODUCTION

The research topic was chosen because of the identification with the object of study, the remnants of quilombos who descend from the black ethnic group and inhabit the territory occupied by their ancestors. When they fled from the slave masters, they sought freedom in distant areas that were difficult to access in order to defend themselves from possible dangers, thus forming spaces for collective coexistence.

The work deals with the issue of quilombo land in the state of Sergipe and the struggle of the Mocambo community to have their territory regulated by the responsible bodies, through institutional procedures. It also represents an alternative for this population to strengthen itself in the face of the possibilities of intense change in the agrarian space.

According to the Federal Constitution, promulgated in 1988, the "remnants of quilombo communities that are occupying their land are recognized as having definitive ownership, and the state must issue them the respective titles". We note that the state must promote access to land ownership, guaranteeing the right provided for in the Constitution.

The process of Brazilian territorial formation and the organization of the agrarian space promoted by the Portuguese crown through colonization, for centuries contributed to the exclusion of blacks and the denial of the territories conquered in the various forms of resistance undertaken.

Throughout history, man has left his mark on geographic space through the movement of each generation. It is in the territory that the realization of life manifests itself, of human existence in everyday life and in its multiple dimensions, mediated by the process of socio-spatial production, appropriation and reproduction, since the historical process materializes in time and space.

It is in the socio-spatial relationships that it becomes possible to understand the constructed space, historically full of contradictions, in which the concrete possibilities for the realization of life are inscribed: man-nature. This relationship must be understood as social production, since, to the extent that man produces nature, he is also its product.

It is from this that everyday practice takes place, in which the plane of the lived is installed and recognized, since the territory also has a concrete practical-sensible dimension and it is in it that society produces and reproduces itself.

It is in space that social relations can be appropriated and built, through the body and its senses. Territory is made up of a dense meaning, given the perception of it as a condition, a means and a product of the action of human beings, and therefore of its use over time.

Through memory, we can make the connection between the present and the past. It is in this sense that a study based on the memories of those who knew, recognized and conceived of the lived space is essential: the place of the historical social subject, of those excluded and silenced throughout official history by the ideological discourse of the dominant power. "[...] memory is not only oppressed because it has been robbed of material supports [...] but because another more damaging and sinister action suffocates remembrance: the official celebratory history whose triumphalism is the victory of the victor trampling on the tradition of the vanquished". (CHAUI, 1994, p. 19).

According to Ecléa Bosi (1994:87), the use of memory is worked on from a perspective in which "Memory is also the possibility of recovering place, revealing another dimension to the space-time relationship". The conclusion is that the final narrative is neither the definitive nor the absolute truth, but narratives articulated by many voices that speak or are also silenced and that generate other voices capable of prolonging the original voice.

In this direction, the method of historical materialism makes it possible to address the problem of conflicts and the territorial planning policy implemented by the state through its institutions.

Historical materialism analyzes contradictions. This method underpins the analysis of human existence, of the relationship between men, relationships that take place through work. In this way, history is understood as procedural and not factual, so dialectics works with cause and effect, but from the point of view of praxis, from the perspective of guaranteeing freedom, happiness, which is only possible when the subject appropriates political power, the result of their relationships.

It is in this direction that we developed this research, the general objective of which was to analyze the process of formation and recognition of quilombo remnants in the municipality of Porto da Folha/SE, in the policy of land regulation of quilombola territories in the state of Sergipe, promoted by the Federal Government through INCRA - the National Institute for Colonization and Agrarian Reform.

To meet this objective, we sought to: a) Recall the history of the remaining quilombo communities in the process of identity; b) Identify the ethnic and cultural elements present in the remaining quilombo communities; c) Identify the economic activities developed in the quilombo communities and how they are articulated; d) Reflect on and verify the importance of public policies in the organization of the quilombo communities.

The aim was to answer the following questions: How do we establish a relationship between land-use planning policies and the ownership of quilombola lands? What is the political discourse on the implementation of territorial policy? How did and does this process of dispossession come about? How do these peoples organize themselves to rebuild their territory?

Observing the main subjects involved in this plot, who have historically fought for the right to land ownership, and who have faced adverse reactions from the public authorities, particularly the legal authorities, which, before guaranteeing the right to quilombola territory, favor the privatization of the land.

In order to develop and advance the research, it was necessary to read up on the concept of Quilombo and the struggle and possession of the land by the remnants. Anthropology is currently discussing and presenting the resemanticized concept, justifying that the previous one does not adequately cover the meaning of what it means to be a quilombola.

Under the new concept, according to Decree 4887/2003, in art. 2, ethnic-racial groups are considered to be remnants of quilombo communities, according to criteria of self-attribution with their own historical trajectory, endowed with specific territorial relations, with a presumption of *black* ancestry *related to* resistance to the historical oppression suffered.

The mapping of quilombos throughout Brazil has provided knowledge of the spatial distribution and location of each territorial unit. In the state of Sergipe, according to the IBGE, this survey indicates the existence of 46 communities and 34 remaining quilombo communities. According to information provided by INCRA, only 14 are in the process of land regulation and are located in the municipalities of Amparo do Sâo Francisco, Aracaju, Barra dos Coqueiros, Brejo Grande, Capela, Canhoba, Cumbe, Frei Paulo, Indiaroba, Japaratuba, Japoatâ, Laranjeiras, Poço Redondo and Santa Luzia do Itanhy. Currently, the Mocambo community, in the municipality of Porto da Folha, has the land regulation title issued by the Palmares Cultural Foundation. According to the location map (Figure 1).

The study area for this dissertation was the Alto Sertâo territory of Sergipe, with the municipality of Porto da Folha as the spatial focus, focusing on the social and political relations of the Mocambo Quilombo community, which inhabits and experiences the common use of land in the same territory.

A survey was carried out of the spatial distribution of the remaining quilombo communities, which have other names: mocambos, rural black communities or "black lands". In order to deepen this analysis, a reflexive reading of the right to possession and ownership of land was privileged, with the aim of understanding the intertwined relationship between the estate of the "black lands" and the process of consolidating capitalism in rural areas, creating a perverse land diversity in which people organize themselves and fight for recognition of the right to land, either through the discourse of territoriality or through a concession from the state, or through popular pressure, in which different social actors fight for a fairer life.

The recovery of identity and the construction of memory has made progress on many issues, however, a number of others related to Afro-Brazilian culture continue to merit reflection, in need of research and knowledge (ANJOS, 1999).

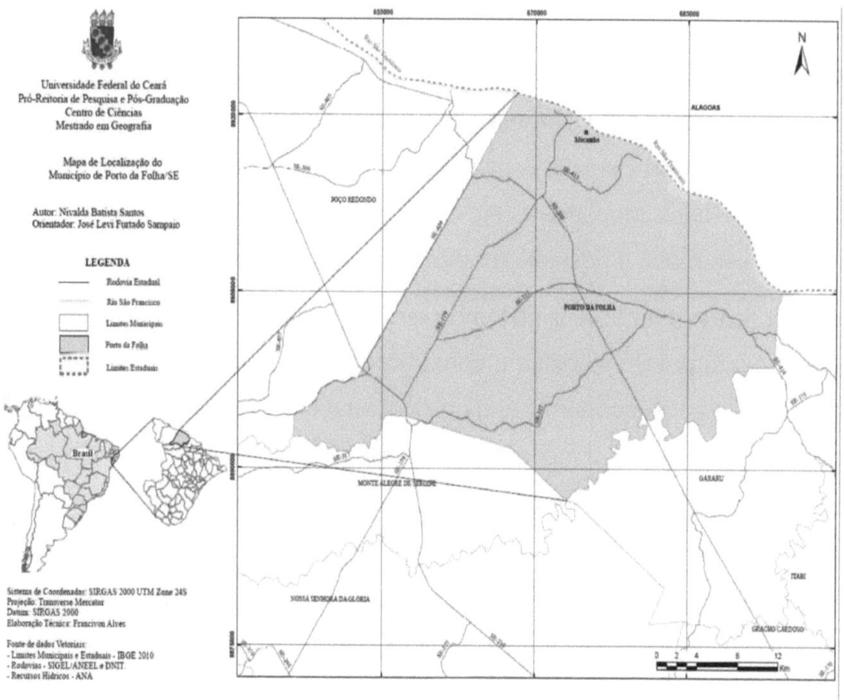

The difficulty of articulating these quilombola communities led to the need to mobilize actions and involve institutional organizations linked to the Black Movement, putting quilombola issues on the agenda in the struggle for recognition of their citizenship rights, according to the Federal Constitution, in art. 68, in the Acts of Transitional Provisions.

As a crucial point, it was important to consider that since Brazil is a country where land concentration is widespread, its agrarian reform model does not guarantee land for work and life for those who live in the countryside or who have been expelled from the cities.

Territory can be represented by society itself in movement, in the continuous struggle for the equally continuous socialization of nature, in opposition to the logic of capital, which produces and reproduces itself in an expanded way in unequal times and spaces, altering the social and territorial division of labour.

The Black Movement has played a significant role in fostering discussion about racism in Brazilian society and in combating the various forms of injustice against black people. As Carneiro (2002) puts it, "in the fight against the various manifestations of racism, sexism and social exclusion". The active presence of the Unified Black Movement and the participation of politicians involved in these causes made it possible for the article to be included in the Constitution.

It is necessary to recognize quilombo remnant communities as a territorial and cultural heritage with significant construction and appreciation of territoriality, based on the socio-economic and political relations present in these territories. For this to happen, it is necessary to change property rights, to change the existence of oligarchies, because land is their source of economic and political power, as Martins (1993) states.

In our view, it is important to identify the changes in the agrarian structure and in production relations in the process of territorial formation in Brazil and in Sergipe, associated with the control of the economic and political elite, which hides the various social practices through an egalitarian discourse for the segments that make up our society.

For the quilombo descendants, the land has other dimensions and appears as a means of production, of peasant re-creation. Peasants and quilombo descendants resist abandoning the land, conceiving it as a territory of freedom and a guarantee of non-subjection to capital.

The geographical dimension makes it possible to deal with the delimitation and regulation of the lands of quilombo remnants in Sergipe and Brazil, pointing out the problems involved in economic, social, political and cultural issues.

Territorial planning policies for quilombola peoples, before meaning access to land, need their identity and cultural legitimization, but the discourses seek strategies that contribute to the subordination of these peoples and reinforce the process of reproducing capital in the countryside.

In the search for a real understanding of the current situation of the remaining quilombo community in question, Mocambo, the main point of departure is the possibility of listening to social subjects who have often been and are forgotten throughout the historical production process, considering these testimonies fundamental for a spatial reading based on memory.

The municipality of Porto da Folha is part of the Alto Sertâo territory of Sergipe, along with six other municipalities: Canindé do Sâo Francisco, Gararu, Monte Alegre, Nossa Senhora da Glória, Nossa Senhora de Lourdes, covering an area of 4,875 km^2 . (Figure 2.)

This territory is located in the semi-arid region with a dry climate, entirely in the basin of the River Sào Francisco and part of the caatinga biome, with a large concentration of settlements scattered throughout the territory. Its territory includes representatives of traditional peoples, such as the Xocós indigenous tribe, located on Sào Pedro Island, and the Mocambo quilombola community, both located on the banks of the Sào Francisco River.

The quilombola community of Mocambo was recognized by the Palmares Culture Foundation on July 14, 2000. It has an area of 2,100.54 hectares, where 130 settled families survive on fishing, hunting, agriculture and social programs such as: Bolsa Escola, Bolsa Familia, fishing insurance and pensions.

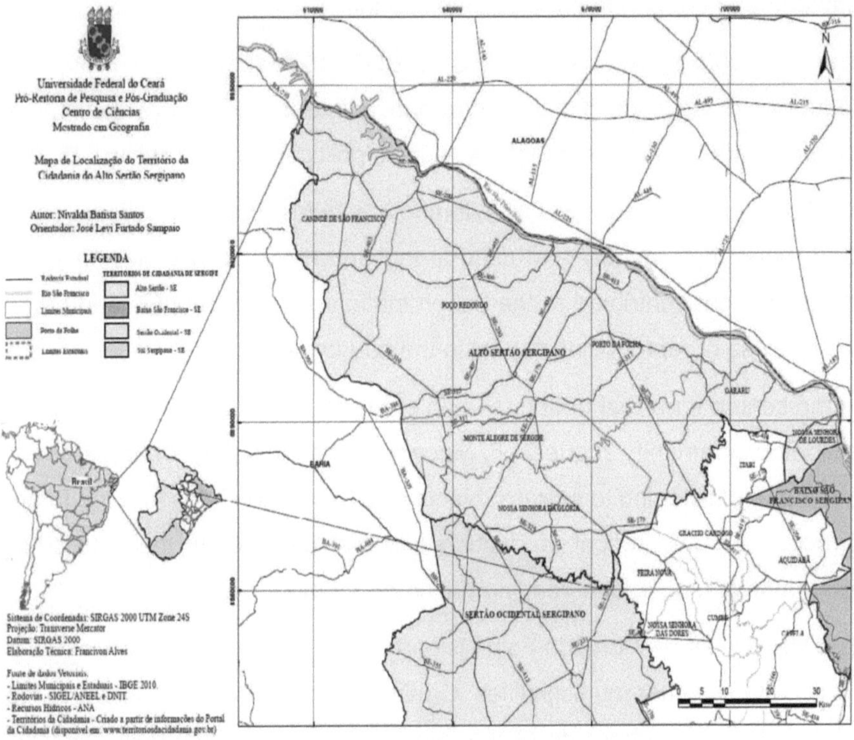

In order to meet the proposed objectives, *on-site* research was carried out using three methodological procedures: observation, description and analysis of the results; documentary research in archives and libraries; and empirical research.

The bibliographical research was based on theorists from different branches of knowledge, with specific reference to readings by: Anjos (1999), Bosi (1994), Bezerra (1984), Moraes (2005; 1994), Martins (1993), Nunes (1991), Conceiçâo (2004), Santos (1978), Silva (2010), Souza (1995), Raffestin (1991), among others.

The selection of study material on the subject was of essential importance to understand the logic established for the recognition of the community, and it was necessary, through bibliographical and documentary surveys, to research books, newspapers, theses, dissertations, periodicals, articles, etc., with the aim of a literary review of the quilombola issue in the local/regional/national context, in an analysis of historical geography, with the specific reality of the Sergipe agrarian space.

Observation was an important procedure for analyzing the black community of Mocambo, requiring a reflexive reading of the reality being researched. Field research was carried out in the territory of the Alto Sertâo sergipano, in particular the municipality of Poço da Folha, with an emphasis on the debate on art. 68 of the Federal Constitution, which sanctions the right of quilombo descendants to access land.

Questionnaires and interview scripts were used with the elderly, adults and young people to verify the historical process of the community which, in the condition of survival and permanence, sought recognition through self-identification as a quilombo remnant, based on a request and the sending of an anthropologist to carry out the RITD (Technical Report on Identification and Delimitation) by the Palmares Cultural Foundation.

The research in archives made it possible to consult documents directly, such as the news in printed newspapers with a wide circulation in the state of Sergipe on the subject, news on websites on the subject, reading publications from scientific events and magazines on the historical-geographical process of territorial formation and on quilombos in Sergipe. Visits were made to public bodies such as: the NPGEO - Post-Graduate Geography Libraries; the BICEN - Central Library - of the Federal University of Sergipe; INCRA-SE; and the NGO CDJBC (Dom José Brandâo de Castro Center).

The methodological procedures are the result of the structure of this final paper, which is divided into four chapters, including the Introduction, with the exception of the Final Considerations. The first chapter deals with the introduction, which discusses the importance of the topic, the objectives, the methodology used and the method on which the research is based.

The analysis was based on quantitative and qualitative information, i.e. the data collected was tabulated and critically analyzed, allowing it to be transformed into qualitative information. As Minayo (2002, p.22) states, "the set of quantitative and qualitative data [...] are not opposed; on the contrary, they complement each other, because the reality they cover interacts dynamically, excluding any dichotomy".

The results obtained are described in this dissertation through the following structure: the first section, which comprises this introduction; the second section, entitled Territorial Formation of Quilombos in Brazil and Sergipe, which explains the process of Brazilian territorial formation, emphasizing the constitution of quilombos in Sergipe territory as a form of resistance to the slave system existing in all regions of the country.

This model of socio-economic structure was established in the colonial period, when blacks were brought in as a productive force to work mainly in the sugar cane and coffee plantations, for the consolidation of the sugar and coffee economy, highlighting the relationship of domination and subservience between the dominators and their dominated.

The third section deals with the quilombola issue, currently on the agenda regarding access to land by the remnants of the quilombo, based on Article 68 of the 1988 Federal Constitution. The concepts of territory, territoriality and identity are discussed as necessary categories of analysis for the recognition of quilombola communities, as well as the legislation that governs the criteria that give descendants the right to access land.

The fourth section discusses the field research carried out in the quilombola community of Mocambo, in Porto da Folha - SE, located on the banks of the River Sâo Francisco, bordering the municipality of Pâo-de-Açùcar, in the state of Alagoas, and its recognition based on the memories of those involved in the struggle for recognition and self-identification, observing the existing contradictions with the public policies in place.

With this discussion, we are betting that its results will contribute to strengthening and understanding the criticism of the contradictions that reside in the debate about the remnants of quilombos, who face the logic of capital and its unequal and combined development, who are not recognized as historical subjects, who in struggle have materialized their territories against the marks of a past that is still alive in the institutional actions of the state, the private sector and civil society in general.

As a result of the research, we were able to conclude that the secular process of colonization of power dominated the social and economic structure of the dominant master-slave relations of the past

"[...] they gain meaning, strength and coherence in this mismatch of the present. The knowledge accumulated throughout a history of social relations and a history of knowledge - which has classified and separated the just from the unjust, right from wrong, good from evil, ours from theirs, the oppressed from the oppressor - is renewed and deepened in the disagreement of modern society [...]". That's why the struggle is most vivid where memory is most vivid (MARTINS, 1989, p. 127).

The reality observed led us to conclude that the struggle for the right to possession and ownership of land was and is fundamental, but that this does not end with the legislative achievement, especially since we can see that agrarian conflicts still persist in Sergipe and in Brazil.

The public policies instituted by the government have failed to guarantee real improvements in the minimum and full lives of the quilombola community of Mocambo, in Porto da Folha - SE, as well as in the entire peasant territory. Land concentration has excluded rural people from guaranteeing access to land in order to ensure the autonomy of peasant family production units.

2 TERRITORIAL FORMATION OF QUILOMBOS IN BRAZIL AND IN SERGIPE

The territorial organization of the state of Sergipe is based on the distribution of land carried out by the Portuguese crown through the sesmarias system, with the aim of promoting colonization in Brazilian territory. In order for the process of occupation and settlement to take place, land was granted through hereditary captaincies.

The territorial formation of the territory of Sergipe was carried out in the same way as the National Project through the Hereditary Captaincies. At the time, it was annexed to the territory of the Captaincy of Bahia, which in the process of spatial planning represented an intermediate area between the captaincies of Bahia and Pernambuco. However, the lands of Sergipe were considered marginal for a certain period and were exposed to piracy on the coast. It was also home to several indigenous tribes who had occupied this territory before colonization.

However, because the land was fertile and vast, it attracted the interest of the grantors who sought to diversify their economic activities. But in this process there were tribes who inhabited the territory, promoting the struggle for the territory that represented the locus of their survival and symbolized their roots, which meant the construction and reconstruction of social relations permeated by actions of solidarity, sharing and collectivity.

During the colonial period, the organization of Sergipe's agrarian space revealed the presence of land concentration, the exploitation of indigenous and slave labour and conflicts between the landowners, the indigenous people and, later, the slaves, who represented the productive force of the prevailing economy, based on the cultivation of sugarcane. The development of this economic activity projected the relations of possession, property and production perceived by the landlord class.

The formation of Sergipe's agrarian territory is strongly materialized by the existence of conflicts over land ownership between the native inhabitants and the Bahian landowners, whose objective was to expand their areas for economic development.

The practices of control and domination were widespread through violence as a way of imposing respect and demonstrating who held power. The situation of exclusion was explicit, both in relation to the indigenous people who had their territories subjected to the domination and control of the hegemonic power of the landlords; and the blacks who came to make up the colonial labor system.

The reality in Brazil, like in Sergipe, after the conflicts over land ownership that took place during the colonial period, created a concentrated land structure consolidated by the political and economic actions of those who controlled and dominated the territory.

2.1 THE FORMATION OF QUILOMBOS IN BRAZIL

The colonial and imperial production system used slave labor in the country's various productive activities, consolidating territorial space and developing alternative ways of obtaining production. Slave labor became dominant in an export economy for the metropolis, which contributed to the expansion of sugar plantations, mainly in the Northeast, contributing heavily to the trafficking of blacks from the African continent to promote the accumulation of mercantile capital.

The Brazilian agrarian space, constituted from the Sesmaria Law, based access to land proportionally to the number of slaves per owner, establishing a land ownership regime that was uncertain and diffuse.

In this way, the relationships established appear as a relationship of conflicts and tensions in which the territories previously occupied by the indigenous people come to be disputed as a strategy to generate colonization and guarantee the exploitation of products. The need for settlement is one of the factors on which the dominant ideology is based for the construction of the country.

It is essential to reflect on the conditions in which black people found themselves and the contradictions that existed at the time, arising from the valorization of space and the importance of the appropriation of territory by the remnants of quilombos.

In a contextualized view, through geographic discourse, it is clear that Brazil's territorial formation conceals social practices that are coated with the character of equality between the various groups that make up society. (Silva, 2010)

These questions should be considered in the light of Geography and History, in which we can look for facts related to the arrival of Africans as a productive force in the colonial economy and the manifestation of opposition to what was being violently imposed, as they were robbed of the right to remain in their original space. Geography can provide the contribution that unravels why these ethnic groups, who contributed to the economic and political formation of Brazil, remain subjugated to the capitalist system.

The analysis of territory must be based on the logic of unequal and combined development through spatial transformations, which impose the need for and adaptation of the valorization of space, while at the same time fragmenting and reifying it. "Territory must be understood as a materiality and as a representation and/or as a project for the practical occupation of portions of the earth's surface and its symbolic legitimation." (MORAES, 2009, p. 29).

The territorial formation of Brazil, during the internal development of colonial society, necessitated decentralization through the establishment of the general government to centralize power to occupy and populate the territory, contributing to the expansion of commercial and mercantile relations, initially with England.

Moraes (2005) understands that territorial formation articulates a dialectic between material construction and symbolic construction, which unites in a movement of economic, political and cultural processes.

The recognition and titling of land belonging to the remnants of quilombos is linked to the process of territorial formation in Brazil and is related to the formation of slavery, which lasted for four centuries from the arrival of the Portuguese colonizers. Colonization can be seen as a process of valuing space, carrying out the stages of appropriation, transformation and appropriation of the space produced.

The Portuguese and Spaniards, in search of new lands to exploit when they found and didn't discover them, because they were already inhabited by natives who, according to the author, were defeated by the astonishment and uneven development of both worlds, immediately sought to discover their riches in order to exploit them. Although in different areas, the technique of exploitation was the same: forced labor, physical aggression, enrichment, colonial oppression.

Therefore, it is necessary to think that space is transformed into territory through the exercise of power and that in the dominant ideological discourse, speeches and proposals are spread throughout the territory as concretization and are constituted as indispensable for economic and political control.

In the appropriation of territory, the conflicts between the various classes that make up society are mediated by the state, which is the instrument used by the dominant classes through laws, decree-laws, decrees and regulations to consolidate the interests of the dominant class, making clear its role as legitimizer of territorial domination.

According to Moraes (2005), territory is seen as the historical result of society's relationship with space, which can only be unraveled by studying its genesis and development. In this sense, the Brazilian territory, based on the slavery mode of production, was founded on the use of the labor force of blacks, captured in African lands, becoming a fundamental component.

They were transformed into slaves and could be found all over the country, working in sugar cane, coffee, cotton and mining, becoming known as the masters' hands and feet for their diverse work activities.

Slave labor shaped Brazilian society during this period, gave it its dominant *ethos*, established the fundamental production relations in the structure of that society and directed the subsequent development of institutions, groups and classes after abolition. (MOURA, 1987: 78)

Faced with this system of slavery, which took place within the territory of a colonial society, which was based on the exploitation of slave labor in various forms, contributing to the expansion of the foreign market, since Brazil was under the domination of colonial relations with Portugal, and in the face of the barbarities imposed on black people who were slaves, there was an urgent need to think about and articulate ways of opposing the system to which they were subjected, since the colonizer's superiority was present at all times, demonstrating who held the power.

The colonizers focused on exploiting the land. The riches, the Indians and the slaves who, when they arrived here, didn't speak the language and, as well as being subjected to forced labor, were forced to become acculturated. The treatment meted out by the landlords led to the emergence of several quilombos in Brazil, in which blacks sought freedom by escaping.

Escapes were a way of rebelling against the process of exploitation and a means of confronting the slave system, seeking ways of resisting the inhuman conditions.

The spatial arrangements that were characterized in the colonial period, through the use of land and the relationships established between men during slavery, contributed to the emergence of quilombos, which produced rural black communities excluded from the current socio-economic context.

The slave was not that passive object who merely observed history, but a permanent dynamic component in the wear and tear on the system, through various forms that acted at various levels in the process of its collapse. (MOURA: 1981, p.8)

The situation of exclusion was explicit, since their cultural rights and territories were subject to the domination and control of the hegemonic power of the landlords. However, "for us, it is precisely in the abandonment of work that the dynamic slave denies the system and asserts himself as a collective historical subject" (MOURA: 1981p. 9).

In these territories of resistance, the blacks carried out activities linked to agriculture and prepared themselves for the confrontation with the armed forces of the imperial government and the troops created to persecute the captives.

The quilombos did not pose any really serious risk to the system. They caused disruption and damage, to be sure, but

they left it intact, configuring an elementary form of struggle and liberation, they proved incapable of subduing and transforming the entire society. (FREITAS, 1982: 45)

In this way, the quilombos represented a way of fighting for freedom and building a model that included collectivity and solidarity, being a means of opposing the slave regime, even if it failed to transform ideological thinking.

According to Moura (1982: 14), there were some forms of organized struggle aimed at achieving their freedom, such as: the organized revolt for the seizure of political power, with the example of the uprising of the Negros Malês, in Bahia; the armed insurrection, such as the one that took place in Maranhâo, with Manuel Balaio as its leader; and the flight to unpopulated areas, which resulted in the formation of quilombos, with Palmares, in Pernambuco, being the most representative.

Quilombos emerged in less accessible places with good visibility of their surroundings, so that they wouldn't be unexpectedly surprised, and were classified according to their characterization and functionality. According to Freitas, (1982) they could be:

Agriculturalists - were found in all regions;

The extractivists - in the Amazon (extraction of the drugs of the sertão);

The miners - mainly located in Goiás, Mato Grosso and Minas Gerais;

The pastoralists - in Rio Grande do Sul there was cattle breeding;

Service workers - blacks left the quilombos to work in urban centers;

The predatory - these lived off the looting practiced mainly against slave masters.

The proposal of the quilombos was the search for freedom and, at the same time, it provided for the establishment of a community in which everyone contributed to activities related to agriculture, carrying out polyculture and integrating the participating members in collective tasks with the well-being of all in mind.

In this way, the organization of the Brazilian territory takes place with the exclusion mainly of the black population, which represents a large percentage of the Brazilian contingent, which masks the concept of freedom. After abolition, blacks were profoundly marginalized, as they were denied access to land and work.

According to Fernandes (1964), the new capitalist relations that developed in 20th century Brazil included black people in a subordinate and subsumed way. The accumulation of capital is expressed in the transformations of the social relations of production and slave labor that were implanted in the country.

The emergence of black territories prevented their owners from getting them back in any way, because they were considered merchandise, and the figure of the captain of the bush appeared to capture them, because the slave, as

Nascimento (1982) states, was devoid of any humanity. They were all victims of patriarchal absolutism, which imposed marginalization and exclusion from the process of Brazilian construction and formation.

In Brazil, they say that a slave needs three PPPs: wood, bread and cloth. And even if they start badly, beginning with the punishment, which is the stick; may God grant that the food and clothing should be so plentiful, as the punishment often is, and with an instrument of great rigor, even when the crimes are certain; which he doesn't use even with animal beasts, some master making more house for a horse than half a day's worth of slaves, because the horse is served and has someone to fetch the grass for him, he has cloth and sweat; saddle and golden bridle. (ANTONIL: 1977, p.55)

The quilombos were formed in areas that were difficult to access, in order to prevent their persecutors from reaching the site. They occurred mainly in the North and Northeast regions, where they had their own organization, principles, values, socialization practices and property regime through collectivism. Their activities favored commercial exchanges between nearby municipalities and neighboring regions, demonstrating that they did not live in isolation.

It was in the Northeast that the Quilombos were very expressive, the most important being Palmares, a territory representative of the Quilombolas' struggle and whose leader was Zumbi, who led the blacks in the battle to maintain the freedom they had built.

The Quilombo of Palmares represented the landmark of resistance against the colonial slave system and assumed an organizational structure in line with that which existed in Africa during the 17th century, in which they lived around a chief who had enough skill to maintain and protect the members of the quilombo.

The quilombo was a constant call, a stimulus, a banner for the black slaves in the neighborhood, a constant call to rebel, to flee into the bush, to fight for freedom. The wars in Palmares and the exploits of the quilombolas took on the character of a legend, something that surpassed the limits of human strength and ingenuity. The blacks outside the quilombo considered Chief Zumbi *immortal* - the flame of resistance against the incursions of the whites. (CARNEIRO, 2011; p.7)

Palmares brought together blacks from various areas of Brazil, from the plantations and urban areas, who carried out activities related to agriculture. Many had a trade and knew how to hunt, fish, raise chickens and other animals.

This quilombo was a symbol of struggle and represented the "Republic of Free Men", due to the large number of blacks and the dynamics in which it was structured. It was not only runaway slaves who took part, but also members of other discriminated ethnic groups and had a very different type of family.

2.2 BLACK RESISTANCE IN SERGIPE

It is of the utmost importance to understand how the process of territorial formation in Brazil took place, and how it relates to the constitution of the space of Sergipe and, consequently, to that of the municipality of Porto da Folha, where the research was carried out. The territorial formation of Sergipe is linked to the process of land distribution, carried out through the donation of land through sesmarias with a maximum extension of three leagues, which were distributed to the chiefs. The territory of Sergipe initially belonged to the captaincy of Bahia and was located between the rivers Sao Francisco and Real.

The death of the bishop of Brazil, Dom Pero Fernandes Sardinha, in 1556 contributed to the persecution of the Caetés Indians, which made it possible for violence to spread to the natives of the Sergipe territory.

As a result of this event, the indigenous people grew to hate the Portuguese and formed an alliance with the French, who were here in search of products from the Atlantic rainforest. This rivalry between the indigenous people and the Portuguese was ended with the intervention of the Jesuits, who thought it was an unjust war that caused the extinction and enslavement of the gentiles.

According to Nunes (2006), the major disagreement between the Portuguese authorities, the settlers and the priests of the Society of Jesus was the latter's intransigence in banning the indigenous people they had segregated in villages or missions under their tutelage. The Jesuits' intention was to catechize the indigenous people within the precepts and dogmas of the Catholic religion, as well as to put an end to the violent action of the colonizers, preventing the initiation of the natives and encouraging them to stay away from the missions.

The scenes of violence and atrocities were encouraged by Luis Brito, but the Jesuits intervened on behalf of those who lived there, calling for an end to the conflict. With the removal of those who promoted the violence, the territory of Sergipe was left at the mercy of the French, who took advantage of it to exploit the brazilwood (NUNES, 2006).

The occupation of the territory of Sergipe was concentrated on the presence of the indigenous people and also on the fighting that took place between the Indians and the Portuguese. The area became known as difficult to access due to the cruelties that occurred during the dispute. At the same time, cattle breeders from Bahia and Pernambuco became interested in the soil, which offered good pastures.

The battle against piracy by the French was a constant threat to the occupation of the territory, which was essential for the colonial economy set up by Portugal. In 1588, the new Governor General of Brazil determined the need to *wage war on the people of the Sergipe coast, punish them and throw them off the land.*

That's why, in 1590, Cristovâo de Barros became responsible for solving the problem of occupying the territory of Sergipe, but for this to happen it was necessary to subdue the natives and expel the French so that they could set up the reorganization of the territory. This happened under pressure from the plantation owners of the Recôncavo Baiano, who wanted to expand their properties, and also due to the expansion of cattle breeding.

During this period, there were several requests for sesmarias and the applicants owned black slaves, who went to work in livestock *farming.* According to Freire (1977), *the people of Sergipe were shepherds before becoming farmers.* The mills began in the 17th century, as the settlers who came from Bahia brought their slaves with them. The Jesuits also encouraged the search for other types of labor found in Africa.

Due to the Jesuits' demands, in 1609, Portugal authorized a charter that absolutely determined the captivity of the indigenous people, whose civilization and catechesis were left to the Jesuits.

Blacks contributed to the process of ethnic and territorial formation of the state of Sergipe, coming from Angola and Guinea. According to Nunes (2006), they were absorbed into the sugar cane plantations and sugar mills, which in Sergipe were located on the banks of the Cotinguiba, Vaza-Barris and Piaui rivers. The estancia settlement was a major recipient of slaves.

The formation of the territory of Sergipe went through various actions in the process of settlement, from the struggles and conflicts with the Indians, to the occupation by the Dutch in 1637, but there is no document that reports the participation of black people. While the Indians allied themselves with the new occupiers, the black slaves took the opportunity to seek their freedom by fleeing to the backlands.

Dutch rule contributed to the emergence of the quilombos known as mocambos here in Sergipe. The formation of these quilombos created a fear among landowners and the authorities of runaways, because the slave was a commodity that they had invested in buying. The formation of the mocambos represented freedom from a system of slavery, which produced consequences through the punishments and torture imposed as a form of control by the master and owner of the plantation.

The emergence of mocambos throughout the state of Sergipe in the 17th century was a form of resistance and protest against the colonial slave system, which viewed them as a productive force devoid of value, but which generated income for the owners and fostered the development of the economy.

One of the first mocambos was Palmares do Rio Itapicuru, but it was destroyed by Potiguares Indians under the coordination of Poti, who received the imprisoned blacks and a sesmaria in Sergipe territory as a reward.

Because of the constant escapes, in the interests of the slave owners, a provision was sent to Brazil *in* which the king of Portugal ordered that *all runaway slaves* who were found should be branded *with an F and*, if they were repeat offenders*, that their ears should be cut off.* This provision was a way of frightening the slaves so that they wouldn't run away and also made it visible to colonial society when the black man had escaped and was recovered and returned to his master.

In Sergipe Del Rey there were three areas where mocambos were formed: 1. the Rio Real, considered to be the most difficult, because it attracted blacks from Bahia; 2. the São Francisco, where the organization of the slaves dismantled the beginning of the colonization of Ilha do Ouro, but in 1698 they were extinguished by Indians tamed by Jesuit action; and 3. the Sâo Francisco. From the forests of Itabaiana, a mountainous area that was difficult to access and concentrated several mills and sugar cane plantations in the fertile valleys of the Cotinguiba River.

The mocambos formed more intensely during the 19th century, when escapes became more frequent and led to a great deal of movement between the owners and the police apparatus, in order to oppress the demonstrations that broke out. Expeditions were planned to rescue the escaped slaves, but the blacks relied on the collaboration of other slaves and free men who provided information, food and weapons. As a result, when the police arrived at the places where they had settled, they found them abandoned.

In Sergipe, according to numerous facts, the quilombolas managed to fight until slavery was practically extinguished. Using clever tactics, sheltering in small clusters of eight to ten houses, which could easily be abandoned at the approach of the repressive troops, and, moreover, maintaining a very efficient system of links with the slaves of the mills, who supplied them with the essential supplies for subsistence, thus not having the need to plant gardens and settle down permanently, these quilombolas wore down the repressive apparatus set up by the authorities of that province enormously. (MOURA, 1988, p. 145)

Escapes were announced in newspapers by describing the characteristics of the runaway slave and offering rewards to those who managed to capture him. There was also the presence of the captain of the bush, whose aim was to go out in search of blacks. As they were considered a fundamental part of the slave-owning mode of production, the owners had no interest in losing them, as they represented a loss to the economy.

In addition to the quilombos, there were other forms of resistance used by black Sergipeans. In 1827 there was the Black Revolt in Sergipe, known as the Nagôs Slave Uprising, in which plantation owners, overseers and administrators were murdered. The movement was suppressed with great difficulty and, as a way of preventing rebellion, laws were created prohibiting them from having access to weapons, carrying knives, wandering the streets at night and gathering in groups.

The rhythm that drove Sergipe's economy was basically linked to sugarcane agriculture, especially in the Lower Cotinguiba region, where there was the greatest concentration of sugar mills and, consequently, a very large number of blacks, in other words, a place where tensions intensified.

2.3 DISCUSSING THE CONCEPT OF QUILOMBO

1988 was the centenary of abolition, which represented the end of slavery in Brazil, but the slaves, when they became free, constituted an excluded workforce with nowhere to stay, as many were expelled from the farms by the owners after having contributed massively to the colonial mode of production.

In that same year, the Federal Constitution (1988) was being promulgated, which, due to the Black Movement, introduced Article 68 of the ADCT (Transitional Constitutional Provisions Act), guaranteeing the right of access to land for the remnants of quilombolas, as it was the will of the public authorities to make reparation for the acts carried out by the society descended from African slaves. However, despite the discussions that were taking place in the country regarding black people, there was no understanding of how the law would be formulated and how it would be accessible to society.

It is essential, however, to understand that the framers of the law did not have enough elements to predict its creative effects. The intention of the legislator, a phantasmagoria often cited in legal hermeneutics texts, can hardly be claimed as the key to understanding this new reality. When we try to give sociological content to this supposed "intention" in the case of art. 68, we find obscure and confused assumptions, a very limited knowledge of the reality that it would represent, as well as a discussion that at no time points to the future, but always to the past (ARRUTI, 2006, p. 67).

In this way, a right was sought within the Constitution. After it was promulgated, society still questions whether the descendants of slaves are really recognized and can ascend to what belongs to them after centuries of exclusion, even after abolition, conceiving a false freedom since they have not been fully integrated into society.

Although this recognition was included in the Constitution, it was almost a decade later that voices began to be heard about a right that had been silenced and had not yet been put into practice.

In 1995, the tricentenary of Zumbi's death was commemorated, and the subject of Quilombos came up in political debates, in academia and in the press, prompting the involvement of social movements in the fight for the rights of the remaining quilombos.

Based on the constitutional text, it was necessary to think about the meaning of the term quilombo, since the one being used was that of the King of Portugal's Overseas Council (1740), which defined it as "any dwelling of more than five escaped blacks, partly destitute, even if they have no raised ranches or pylons". To be considered a quilombo, it had to have the characteristics described as a form of reaction and resistance.

The quilombos in colonial society had to be extinguished in order to prevent them from strengthening and confronting each other and gaining control of the colony, since there was a large percentage of slaves in the population.

After the abolition of slavery, after centuries of submission, the black man was relegated and deprived of value, no longer fit to work, because in this relationship of dependence he became invisible in the eyes of the landowners.

> For the slave, skill is assumed adaptation or maladaptation, the cunning that makes him live, the cleverness born of experience and forged in adversity, which will mark all Brazilian life, and that of slave men as well as freed men and even free men. Thanks to his way of knowing how to live, the slave strives to hasten the passage that leads to the mythical, lost past, through the difficult present, to an idealized future of freedom. In day-to-day life, the way makes it possible, because linking past and future is knowing how to survive. This is what unlocks imaginable treasures, the riches of slave life. Bought, sold, sent away, the slave knows how to preserve his share of autonomy, humor, tenderness and dreams (MATTOSO, 1982, pp. 174-175).

Blacks, who were a productive force in the sugar cane fields and on the coffee plantations, were no longer useful in boosting the economy, which focused on free labor and the arrival of immigrants to replace slave labor. Deprecating images are created and they come to be seen in a negative light.

According to Ianni (1988:236), the image is that "[...] the black man is active and hardworking when he is a slave, when he is not, on the contrary, he becomes corrupt, a drunkard, a libertine and a thief". In other words, the image passed on by society is aimed at denying access to land by imposing degrading characteristics on it.

The material and symbolic repression against black people favors the struggle for territories conquered during the period when quilombos were formed or when they remained on the properties of their former masters after they went bankrupt. According to Azerêdo (1987:27), the "other" is someone who is strange, different not only in appearance, but also in values, beliefs, lifestyle and social position.

Nina Rodrigues (1977) sees the quilombo as a form of African persistence in Brazil, a return to barbarism. Due to the large contingent of blacks making up the population structure, they were used as a productive force and their reaction to the mistreatment they were subjected to resulted in the use of violence.

For Carneiro (1988), the quilombo was a counter-cultural phenomenon, a rebellion against the standards of living imposed by official society and a restoration of old values. Due to the imposition of the habits and customs that existed in the colony and the prohibition of the realization of their culture by the lords and landlords.

For Nascimento (1980), quilombo does not mean fugitive. Quilombo means fraternal and free gathering, solidarity, coexistence, existential communion. For them, it was being together, dividing up activities, discussing ways of solving problems and recovering the sense of African communitarianism that many knew so well.

In 1981, the Catholic Church, seeking to redeem itself from the past, held the Quilombo Mass, acknowledging its guilt for having contributed to the persecution of blacks and quilombolas. This act opens up the question of what it means to be a quilombola, not in the past, but today (HOORNAERT, 1982, p.12).

There are various conceptualizations of the concept of quilombo and the discussion also includes the participation of the academic debate, not just the views of official technicians. In this sense, other definitions are proposed, such as Moura's (1994) formulation of contemporary quilombos, characterizing the communities:

Rural black communities that bring together descendants of slaves [who] live off subsistence farming and where cultural manifestations have a strong link to the ancestral past. This link with the past has been rectified, chosen by the inhabitants as a way of maintaining their identity (MOURA, 1993, p.28).

The Brazilian Anthropology Association (ABA), through its Working Group on Rural Black Communities, has also proposed a new resemanticized concept of the term quilombo as groups that have developed practices of resistance in order to maintain and reproduce their characteristic ways of life in a given place, whose identity is defined by a "common historical reference, built on shared experiences and values". This concept will make it possible for another community with the same concept to access the law. (GT ABA, 1994)

The ABA Working Group, Terra de Quilombo (Quilombo Land), formed in 1994 in the face of a whole debate on the concept of quilombo, in 1999 opposed the preliminary draft proposed by the federal government, which went backwards and curtailed what had already been discussed about the term quilombo. According to this draft decree, the occupation of the territory had to date back to before the abolition of slavery, which went against what was established in the Federal Constitution, thus usurping rights that had already been conceived. This document was written during the 6th Meeting of the Provisional National Commission for the Articulation of Quilombos, held in Sâo Luis/MA.

For the ABA, anthropological production must be based on three parameters: 1) understanding the foundation and objective of the demand; 2) clearly defining the methodological procedures and technical support that guide the production of the document, acting as the canons of anthropology; and 3) proceeding with the argumentative exposition with objectivity, paying attention to the objectives of the document, resorting to ethnographic demonstration in the most pertinent way to these ends.

Part of the concept that was discussed by the ABA and which is complemented by Decree 4.887 of November 20, 2003. Art. 2 - For the purposes of this Decree, the remaining quilombo communities are considered to be ethnic-racial groups, according to criteria of self-attribution with their own trajectory, endowed with specific territorial relations with a presumption of black ancestry related to resistance to the historical oppression suffered.

> [...] the term "remanescente" introduces an important difference from the other use of the term quilombo in the 1988 Brazilian Constitution. In it, what is at stake are no longer the "reminiscences" of former quilombos (documents, remains of slave quarters, emblematic sites such as the Serra da Barriga, etc.) of Article 215 (Section II "On Culture"), but "communities", i.e. social organizations, groups of people who "are occupying their lands", as "Article 68" says. More than that, it concerns, in practice, groups that are organizing themselves politically to guarantee these rights, and therefore claiming this designation from the state. Therefore, what is at stake in any collective effort for official recognition as a remaining quilombo community is always (so far) the land conflicts in which these communities are involved, and not any memorialistic desire to assert themselves as continuities of those metaphors of slave resistance and the "African world among us", which were the historical quilombos (ARRUTI, 2006, p.81-82).

In the political context, legitimization demonstrates that the dominant discourse remains strong, contributing to the transgression of the norms defined by the 1988 Federal Constitution, which guarantees ownership only if the state issues the title. But the discourse of legitimization discursively defines how this social and, at the same time, dominated group finds it difficult to fight for their acquired right. This process affects social representations, and this discourse explains the reproduction of ideologies.

In this way, the dominant group practices, reproduces or identifies with the discourse of legitimization. And the dominated group suffers the violation of values and the naturalization of exclusion because it has a history of devaluing its ethnicity. However, in the globalized world, the emphasis is on the economic aspect, with the liberal and mercantile ideology predominating, and the social aspect being relegated to actions developed by civil society.

Therefore, in Brazil, it was necessary for social movements to demand that the conquests be fulfilled, because in the institutional sphere, the state has its interests focused on defending the dominant group, perpetuating the reproduction of power, which is based on the veiled discourse of guilt against social actors who lack political knowledge.

3 THE QUILOMBOLA QUESTION

The permanence of power relations in the countryside, established from colonization to the present day, contributes to the dispute over territories between the parties, but the land for peasants fulfils a social function and becomes an object of work, survival and resistance.

In Sergipe, one of the forms of resistance is the struggle to remain on the land, where the squatters believe in their right to the territory appropriated by their relatives and ancestors, who have built an affective relationship with the land and who, through their crops, make a living, live and produce for the market.

3.1 TERRITORY AND IDENTITY

These concepts are fundamental, but they undergo a great deal of modification in their historical retrospective, seeking to understand the processes of building social and economic relations, observing the different eras.

Geography seeks to understand that the notion of territory becomes essential to understanding what is being produced by the black community, the locus of the research, and how relations are subordinated to the mode of production with all its conflicts, contradictions and mediations.

From this discussion, we seek to understand the struggles of the remaining quilombolas for land ownership, since they were squatters, while art.68 gives them the possibility of legally owning the land through their ancestry.

However, it should be borne in mind that this concept comes from the natural sciences, mainly due to the influence of Darwin's and Lamarck's theories on this author's works. The idea of territory is linked to the appropriation of space as a form of struggle for land.

In this sense, we emphasize the studies developed by Raffestin, going back to Ratzel, with regard to the relationship between territory and state. This author also points out the great confusion caused by geographers in using the concepts of space and territory as equivalent terms. In this way, space is prior to territory.

Territory (...) is a space in which work has been projected, be it energy or information, and which consequently reveals relationships marked by power. Space is the "original prison", territory is the prison that men build for themselves (RAFFESTIN, 1993, p. 143-144).

Thus, in support of the political nature of the notion of territory, Raffestin also points out that the territorial image projected by a social actor does not correspond to the real territory, since this is the combination of different territorial projects in dispute. Territoriality can only be understood in its socio-historical and space-time context.

For Andrade (1994), the concept of territory should also not be confused with that of space, so "the idea of territory should always be linked to the idea of power, whether we are referring to public, state power, or the power of large companies that extend their reach over large territorial areas".

However, "territory has a political and affective dimension, or both". In the political dimension, this concept is associated with effective control by institutions or groups over a given segment of space. On the affective side, it derives from the specialized practices of different groups, highlighting the symbolic issue, the feeling linked to the appropriation of the territory.

For Souza (1995), territory can be defined as a field of forces, a web or network of social relations that defines a boundary and an otherness, contrasting those who are part of and those who are foreign to that space.

In this way, "territory (...) is fundamentally defined and delimited by and from relations of power" (SOUZA: 1995). For this reason, he considers that territory has usually been associated with the nation state, but points out the need to analyze it on its various scales, considering the possibility of various powers acting on the same territory.

Moreira (1996) sees space as a given on which men organize territory. For

Moreira, this is a Kantian vision, which is therefore incapable of dealing with current spatial processes.

For Moreira, territory is a spatial expression of organization with respect to domain, constituting a fraction of space, a domain within space, a project of territorial construction. Territorialization, on the other hand, is defined as a process of fixation, of territorial, environmental and cultural rooting; also considering the new territorialities, resulting from space organized into networks, with no longer any link, therefore, between territory and the environment, complementing Badie's (1996) analysis, which focuses on the transnational identifications referred to economic, mercantile, cultural, religious, migratory or professional networks that tend to override the principle of territoriality. For the author, these contradictory processes produce and reproduce themselves in different parts of today's world, creating processes and relationships of interdependence between states, nations and companies in different parts of the world (OLIVEIRA, 1998, p. 08). From then on, he emphasizes the importance of studying territory, which he believes must be understood:

> (...) as a contradictory synthesis, as the concrete totality of the process/mode of production/distribution/circulation/consumption and its supra-structural articulations and mediations (political, ideological, symbolic, etc.) where the state plays the role of regulator. Territory is thus a concrete product of the class struggle waged in society in the process of producing its existence.

At the same time as we talk about territorialities, we can also talk about deterritorialities and reterritorialities. Business strategies, the logic of mobility, migration and the exchange of information and knowledge are all aspects to be considered as promoters of deterritoriality. In this way, we seek to understand how this struggle has been taking place in the semi-arid region of Sergipe, in order to verify how the territorialization of black people in this area is taking place, in the implementation of this right. In order to do this, it is necessary to take a historical look at Brazil's agrarian reality, where, based on its development and contradictions, movements to fight for land ownership have become a reality.

The ideological formation was based on a knowledge of the territory founded on the politics of consolidation and appropriation, since the elite had the competence to carry out the politics of favor. In this way, in Sergipe, the areas of conflict over land ownership, especially in the semi-arid part of the state, pushed the remaining quilombo peasants to seek and demand the right they had won under Article 68 of the Federal Constitution of the Transitional Provisions Act.

In this sense, it demonstrates the role of the nation-state and the new functions it has taken on as a result of the globalization process, highlighting the interdependence of places, with territory being the basis of the nation-state.

> Today, when we live a dialectic of the concrete world, we have evolved from the notion (...) of the territorial state to the post-modern notion of the transnationalization of territory. (...) But just as before everything was not, so to speak, "nationalized" territory, today everything is not strictly "transnationalized". "Even in places where the values of globalization are more operative and effective, the inhabited territory creates new synergies and ends up imposing a revenge on the world". (SANTOS, 1996, p. 15).

At this point, the author points to the possibility of solidarity happening, based on the creation of new solidarities between people and places.

> It's a conflict between banal space, a space experienced by all the neighbors, and global space, inhabited by a rationalizing process and ideological content from distant origins that arrive in each place with objects and norms established to serve them. "Hence the interest in returning to the notion of banal space, i.e. everyone's territory, often contained within the limits of everyone's work; and in countering this notion of networks, i.e. the territory of those forms and norms at the service of a few" (SANTOS: 1996, p. 18).

Santos draws attention to the force of the market, which even cuts across people's interests, from the political side of this perverse globalization, i.e. market democracy and neoliberalism, which imposes its logic on the various territories. However, these do not reduce the possibilities of ways of living based on solidarity, contiguity and neighborhood solidarity.

In another approach, Silva and Silva (2003, p.10) point out that:

Territories not only have the characteristic of classically expressing complex power relations on different scales, and no longer only on a national scale, but also the perspective of valuing their identity, their bonds of cohesion and solidarity (...) the tendency to organize themselves socially and politically towards development that is effectively sustained from both an economic and social point of view, as well as from an environmental perspective.

In addition, historical materialism cannot lose sight of the fact that the contradictions engendered by the capitalist mode of production do not allow us to consider the real possibilities of "sustainable development" from a socio-economic point of view, since we consider that the interests of capital are contrary to the interests of society as a whole.

It is believed that the interests of the World System, and even the interests of the state, do not correspond to the interests of society as a whole, since this produces inequality. We therefore conclude that territory, as a political dimension, is represented as a force field of divergent interests.

In this context, the importance of the struggle by a section of excluded society to participate, even if only partially, in the production of the territory, also imprinting its territoriality, is highlighted.

These contradictions can be seen throughout the Brazilian countryside. However, in opposition to the consolidation of the logic of landlordism and the appropriation of the countryside by capital, various initiatives and even national resistance have emerged in various parts of the country, mainly from rural social movements, especially from the 1980s onwards, with the end of the military dictatorship and the return to "democracy".

In this scenario, the struggle implemented by the black movement for the recognition of rural black communities throughout Brazil stands out, organizing itself to pressure the state to recognize a right provided for in the constitution through Article 68 of the 1988 Federal Constitution.

For Castells (1999, p.93), "with the exception of an elite few globopolitans (half human beings, half flows), people all over the world resent the loss of control over their own lives, their environment, their jobs, their economies, their governments, their countries and, ultimately, over the planet".

However, despite this unfavorable scenario for the majority of the world's population, the author also points out that: "resistance confronts domination, empowerment reacts against disempowerment, and alternative projects challenge the logic inherent in the new global order, increasingly perceived by people across the planet as disorder." (p. 94)

Thus, the essence of the contradictions engendered in the current system makes it essential to look to the studies carried out by Oliveira (1998), when he seeks to demonstrate the contradictions inherent in the process of capitalist development in the countryside.

(...) The study of Brazilian agriculture must take into account that the process of development of the capitalist mode of production in the Brazilian territory is contradictory and combined (...) At the same time that this development advances producing specifically capitalist relations (implanting salaried work through the presence in the countryside of the "bóia-fria"), it (capitalism) also produces, and equally and contradictorily, peasant relations of production (through the presence and increase of family work in the countryside). (p. 09)

However, it is this contradiction that gives rise to the historical possibilities of understanding the different and unequal territorial formations and regions as concrete territorialities, historical totalities, therefore, of the contradictory spatialization of capital (production/expanded production) and its links with landed property, i.e. land (OLIVEIRA, 1998, p. 09). It is in the territory that representations materialize in physical space and also where discourses are incorporated.

According to Silva (2004, p.74), "To speak of contemporary territories is to speak of spaces where power is exercised. The territory in which the black community establishes its working relationships and group coexistence, exercising its territoriality through the links they establish in the human-nature relationship."

The colonial production system absorbed slave labor in the various productive activities carried out in the country, consolidating the territorial space, which could be occupied by groups, and developed alternative forms of production. The social practices of control and domination will be preserved in the spaces even after the abolition of slavery.

The situation of exclusion was explicit, since their cultural rights and territories were subject to the domination and control of the hegemonic power of the landlords. The conflicts waged in defense of their rights led to a questioning of the actions and role of the state, as well as the concept of property. For peasants, land is life, it is a space where they are guaranteed their permanence, survival and the mark of their ancestry.

The indifference of the state to the land situation of the remaining quilombolas, this category which, because they don't have the deed to their land, provides them with the conditions to regulate their territory and, consequently, access to technical assistance and credit to invest in their production.

The aim is to understand how this struggle is taking place in Sergipe, in order to see how the remnant community, through its territorialization, is resisting the expropriation of its land. In a reflective reading of land rights and land tenure, it is necessary to analyse the imbrication in the process of consolidating capitalism in the countryside, creating a perverse land diversity among traditional peoples and quilombo remnants, where they organize themselves and fight for recognition of the right to land, either through the discourse of territoriality by the state or by popular pressure in different subjects fighting for a better quality of life.

In the municipality of Porto da Folha, in 1992, as soon as the remnants recognized themselves as quilombolas, a conflict began, provoked by the landowner who destroyed all the plantations, demonstrating his power through force to intimidate them.

The big landowners, who were used to being in control and taking control of their destinies, feel threatened by the strengthening of the group of quilombo descendants, who are seeking to realize their desire to own the land.

But the contradiction between the social function of land for the quilombo remnants, the category under study, and for capital is reinforced by the state, which legitimizes landlordism, responsible for economic backwardness and the great exclusion of rural workers.

This same state is indifferent to the power relations that take place in these territories of resistance, which, linked to the judiciary, ignores the legitimate right of this group and defends the right to land ownership as an absolute and unlimited right of a minority that is the ruling class.

What stands out is the knowledge of the forms of power and domination that permeate the conflicts in the various territories that are established in the agrarian space in Brazil and Sergipe, where the subjection of the other takes away their right to recognition of their historical role.

For the quilombo descendants, the land has other dimensions and appears as a means of subsistence, which is why they resist abandoning the land, seeing it as a space of freedom and a guarantee of not being subject to capital.

The land policy that is being developed will allow us to understand the ordering and reordering of the territory in the face of the struggle for land ownership by traditional peoples.

Perhaps we can build a new historical moment, but this requires changes in the socio-cultural, political and economic context, no longer centered on exploitation and domination, but on the emancipation of the classes as subjects.

The relationship between the territory and the remnants of quilombos is responsible for the community's name, which is eminently from the African dialect, and the older descendants say that it was inhabited by blacks who lived by fishing in the river and growing rice to ensure their survival. They had a peaceful relationship with the owners of the farms, until one of the owners decided to destroy the entire rice plantation in retaliation, because they wanted them to leave the land.

This is when the whole process of fighting for land ownership began, no longer as squatters, but as remnants of quilombos, and for this to happen they had to self-identify as this category in order to have access to what was described in art. 68 of the Federal Constitution. And the territory, based on anthropology, would be linked to the past through oral history and memory as a way of recovering the social and cultural relationships experienced by their ancestors, which would have been passed down from generation to generation.

The affirmation that they have always been linked to the land as their main source of survival and that it is essential to the process of constitution of the community, because it is both peasant and quilombola, since the land represents a natural, finite asset that cannot be created by labor, but can be appropriated and dominated by the capitalist system.

3.2 The Earth

The debate on land ownership in the area of quilombo remnants plays a very important role, since they survive on what is produced and characterizes their identity. It is a source of memories of their ancestors and has a historical value based on the relationships that are established on a daily basis. In this sense, the inseparable political and social problem must seek solutions to the pressures exerted by the ruling class on land use, and it is necessary to propose results for the current demand that will express the hegemonic interests of the period.

The process of man's socialization, as he settled down and sought the means to survive from the time he was a hunter to the time he became a farmer and shepherd, growing crops and raising animals to meet his needs. Living in a community and in a collective production system, this led to a division of labor by age or strength in carrying out activities. However, the families came together to form villages, realizing the need for security and protection.

In this way, man began to select the seeds, producing more than he needed, forming a food surplus, leaving some primary activities and dedicating himself to others, creating a new social division of labor, thus giving rise to social institutions within the structure of relations of domination and exploitation.

From a perspective where the history of man and his relationship with nature is unquestionable, whether on any scale or according to the relationships established, since it is taken as a resource to be exploited and mediated to meet needs.

When historically the situation materializes in the face of political, economic and often cultural fragility, using means to subjugate others in the name of their own well-being, one can see a Brazilian society with a marked fragility in terms of collective conscience.

The political and hegemonic groups have remained in power through force and coercion, which allows them to be easily manipulated and subjected to the tyranny and power of capital, since the country is divided between those who own land and those who don't, when we talk about the countryside.

The land structure in Brazil is based on latifundia. Although this concentration of land initially existed, there was not yet an agrarian question, which was established in the 19th century with the advance of capital in the countryside, promoting a change in labor and production relations, in which land ownership is denied to some, but is a condition for others to legitimize power, which through discourse tries to promote the disintegration of the peasant in its essence and removing the possibilities of appropriation as it has control.

In this situation, new forms of struggle are being recreated in the face of the conflict that has arisen from the perception and awareness of those who are being excluded from the process and exploited, since for peasants the loss of land limits their living conditions.

In Brazil, the land structure, initially centered on the sesmarias, represented donations of land to people who were able to exploit it in order to promote the settlement of extensive areas absorbed by the Portuguese crown, whose objective was to protect the territory from invasion by other nations.

However, in the past, settlers only had access to land through ownership and if the land was cultivated. But in the colonial period, land began to have a market value after the Land Law and, according to Smith (1990), it was intended to meet the objectives of the ruling class, mainly in the Center-South.

The purchase and sale of land in Brazil led to the concentration of land, to which settlers and blacks had no access, since its value as a commodity ruled out this possibility. The Land Law and the abolition imposed were aimed at the interests of England. In this way, it promoted the capitalist process and eliminated the slave system.

According to Smith (1990), the Land Law should be understood as a way for the state to have control over wastelands and establishes some points, such as: the prohibition of wastelands, except by purchase; penalty and eviction for those who take possession of wastelands; the government starts to measure and demarcate to separate public and private domain; it creates the land registry that must be done within the stipulated deadlines.

Various conditions are imposed by the law, which prevents the dispossessed from buying land, which also has an exchange value, leading to unequal spatial production in the countryside, since the way of life of the excluded will not make ownership possible.

The peasant seeks fulfillment through land ownership. For Marx, man realizes himself and appropriates what he takes from nature to meet his needs, which is why the remnants are fighting for land titles, through the relationships of belonging that have been established since their ancestors, observing use value and exchange value.

According to Conceiçâo (1991), land is the guarantee, it is the certainty, it is the will, the dream of change, the denial of exploitation and submission. In other words, for decades, families continued to serve the landowners with their labor power, being sharecroppers and tenants subject to the interests of others.

When analyzing the contradictions in the countryside regarding the land question, which are associated with the capital question, Oliveira (2001) says that the expropriation of land in the country opens up historical possibilities of return, through the struggle for possession/ownership waged by expropriated peasants or those who have never had access to land in Brazil.

Peasants are marginalized, but they don't disappear, they find a way to resist in the countryside. Even though they are expropriated, the struggle for possession becomes fundamental to their existence. To define the peasant, Moura (1988) cites the peasantry as an oppressed pole of any society. In any time and place, the peasant's position is marked by subordination to the owners of the land and power, who extract different types of income from him: income from produce, income from work, income from money.

It is important to note that the peasant's permanence on the land and his relationship with the owners of private property transform him into a partner, tenant farmer and sharecropper, subjecting him to the production process of capital.

3.3 LEGISLATION ON QUILOMBO REMNANTS

The political discourse on the right acquired by art. 68 of the Federal Constitution promotes discussion of the land question and the historical recognition of the contribution of black people in the process of building the country, mainly through the economy as a productive force.

This discussion is causing a reaction among landowners who are part of national politics or who have representation in the Chamber of Deputies and the Senate.

"This structure, made up of fundamental, socially shared cognitions related to the interests of a group and its members, is acquired, confirmed or altered mainly through communication and discourse". (VAN DIJK: 2006 p.43).

So, we can analyze political discourse through legitimation and political cognition, concepts worked on by Van Dijk (2006), who discusses legitimation as an important role in the use of language and discourse, even as a social or political act carried out through text and conversation.

Considered from a pragmatic approach, legitimization is related to the speech act of self-defense, i.e. the political or social actor who proves for acceptable reasons, grounds or motivations that their once criticized practices are pertinent to a legal context. As such, it takes place through a complex communicative act involving a set of interrelated discourses (Van Dijk, 2006, p. 318-319).

In this way, discourses are legitimized through their institutional nature and not only apply to political and social actors, but also to institutions and society. Legitimizing discourses presuppose norms and values, and explicitly or implicitly determine what is "correct" within the political and social context in general (Van Dijk, 2006, p. 319-320).

In discourse, we can see through the mechanisms used by the ruling class, who hold power, control political and social representations according to their interests, demonstrating power, but contributing to the reproduction of power through discourse. According to Van Dijk, "control applies not only to discourse as a social practice, but also to the minds of those being controlled, i.e. their knowledge, opinions, attitudes, ideologies, as well as other personal or social representations." (p.18).

For this reason, the relationship between discourse and social power considers that power exercises a form of social control if it is made up of socially prominent resources. In this way, Van Dijk repeatedly warns of the importance of the ideological structure for the exercise and preservation of social power, encompassing fundamental, socially shared cognitions based on the interests of a group and its representatives.

The author emphasizes that the solution to this impasse can only be conceived by *counter-power*, as a form of resistance to constituted power. Looking at the history of humanity, we can highlight the ability of human beings to react by being aware of themselves, of the sphere to which they belong and of the importance of their role in the structure of the society they are part of. In this way, they can promote change in the *status quo* and rebuild trust and be managers of their world.

3.3.1 Article 68 of the ADCT of the 1988 Federal Constitution

The history of the formation and constitution of Brazil imposes on us a different reality in which an ideological discourse is necessary for the maintenance of dominance, where ideology serves as the basis for the permanence of certain groups considered dominant. "If controlling discourse is a major form of power, controlling people's minds is another fundamental way of reproducing dominance" (Van Dijk, 2005, p.26).

Discursive strategies will contribute to social and political representations transforming relevant issues into something unimportant in which the state remains indifferent, breaking the rules when it doesn't enforce what the article says.

However, political practice can be legitimized in the political context. It is clear that the hegemony of the current system imposes its determinations based on the economic conception, without taking into account the individual and their social memory.

The use of the expression "definitive possession is recognized" makes the state responsible for legalization, given the veracity of the fact stated in the Magna Carta, a reliable source of reference for the country determining that it be executed.

As for the quilombo remnants and their representation as social actors in the process of struggle for land ownership, it is of paramount importance to identify them in order to legitimize their right as described in Article 68 of the Civil Code.

Recognition takes place when Decree 4887/2003 determines the meaning of quilombo remnant, described in art. 2: "For the purposes of this decree, the remnants of quilombo communities are considered to be ethnic racial groups, according to self-attribution criteria, with their own trajectory endowed with specific territorial relations with presumption of black ancestry related to the historical oppression suffered".

We can see that identification is not just a discursive process, but the construction of meaning when the text talks about "the criteria for self-attribution and the presumption of black ancestry", the actors internalize and assume their identity can be transformative.

According to Ramalho and Resende:

> [...] in relation to the study of the materialization of styles in texts are relevant themes: the internalization of discourses in the construction of identities and identifications; the relationship between identity constructions and modes of discursive and social action; identifications constructed for specific groups in widely circulated texts or in various types of texts; identity construction in ethnographic research and life stories; the relationship between identities and social movements or other groups linked to specific struggles. (2011: 72)

In an interview with Revista de Direito Agrario (2007), the Deputy Prosecutor of the Republic, Deborah Duprat, answered the question of land ownership, whether there were differences between the demarcation of indigenous peoples' lands and the titling of quilombolas. She replied: "I would say that, from the ethnic, cultural and legal perspectives, there are none."

We have the assertion that the process of recognition stalls due to the decision, which depends on the dominant group in power representing its own interests to the detriment of those of others. Even when the dominated group argues that it is in the interests of the community and of its representation, it is predisposed to value itself based on its identification as a remnant, in order to convince that its struggle is essential to obstruct the prevailing discourse.

According to Fairclough:

> [...] the selection of previous texts and types of texts articulated in a given instance (a particular "discursive event") and the way in which they are articulated depend on how the discursive event is situated in relation to hegemonies and hegemonic struggles. (2008, p. 29)

Since the state itself favors the non-execution of the issue

This process also includes the exclusion of the black ethnic group, even though they have contributed to the economic and territorial formation of Brazil.

> In situations, ordinary people are passive targets, to a greater or lesser degree, for example, of their bosses or authorities [...] who can simply tell them what they should (or shouldn't) believe or what they can (or can't) do. (VAN DIJK; 2008 P. 119).

The interview poses another question to the Deputy Prosecutor: are you concerned about the negative aspect that appears in the legal opinion that the groups that could be recognized had to prove that they were quilombos at a certain time? (Revista Direito Agràrio: 2007).

The use of the term "residue" represents the leftover or remainder of something; this term determines a process of exclusion of people and groups who are of black ethnicity and who, in the eyes of this hegemonic society, have no value. But the word "absurd" used by the interviewee shows her dissatisfaction with those who believe in and legitimize segregation/exclusion and promote inclusion.

So how do we organize the territory and distribute the land to the quilombo remnants on the basis of the national land-use planning policy? Why, in order to organize, is it necessary to include in time and space all human, industrial and agricultural activities, their forms of occupation and the production of goods and services? Why organize? The aim is to promote economic development, quality of life, the management of water resources and the rational use of the territory. But for this to happen, it is necessary to resolve issues such as the social inequalities that are present throughout the country, including access to land for the remnants of quilombos. Therefore, it is essential that the state apparatus pays the utmost attention to the claims and legitimizes the liquid and certain right present in the Constitution.

Territorial planning, based on the premise of land ownership through the identifiable representation of the subjects, will give a new configuration to space through the territorialization of those groups considered to be dominated by the dominant hegemony.

In the political context, legitimization shows that the discourse of the dominant ideology remains strong, contributing to the transgression of the norms defined by the 1988 Federal Constitution. However, the discourse of legitimization discursively defines how this social and at the same time dominated group finds it difficult to fight for the right it has acquired. This

process affects social representations, and this discourse explains the reproduction of ideologies.

In this way, the dominant group practices, reproduces or identifies with the discourse of legitimization and the dominated group suffers the violation of values and the naturalization of exclusion, because it has a history of devaluing its ethnicity.

However, in a globalized world, the emphasis is on the economic aspect, with the liberal and mercantile ideology predominating, and the social aspect being relegated to actions developed by civil society. Therefore, in Brazil, it was necessary for social movements to take action to demand that these achievements be fulfilled, because in the institutional sphere, the state has its interests focused on defending the dominant group, perpetuating the reproduction of power that is based on a veiled discourse of guilt against social actors who lack political knowledge.

The analysis shows that political practice is a discursive practice that uses strategies to produce its legitimacy, in which the institutions that hold power over the dominated groups generate the abuse of power.

3.3.2 Decree 4.887/2003

Decree 4887/2003 regulated the procedures for the identification, delimitation, demarcation and securitization of the lands occupied by the remnants of quilombola communities referred to in art. 68, ADCT.

According to Article 2, the concept of remaining quilombo communities requires quilombo identity, in which case it is necessary to belong to "ethnic-racial groups, according to self-attribution criteria, with their own historical trajectory, endowed with specific territorial relations, with presumption of black ancestry related to resistance to the historical oppression suffered". The land they occupy must be used for their physical, social and cultural reproduction.

In order for the land to be measured and demarcated, the territoriality of the community must be taken into account, and it is possible to present the documents for procedural instruction. It is the responsibility of the Ministry of Agrarian Development (MDA), through INCRA (Institute of Colonization and Agrarian Reform), to identify, recognize, delimit, demarcate and title the lands occupied by quilombo remnant communities, without prejudice to the concurrent competence of the States, Federal District and Municipalities.

INCRA is responsible for regulating administrative procedures and may establish covenants, agreements and similar instruments with federal, state, municipal and Federal District public administration bodies, non-governmental organizations and private entities.

The administrative procedure must be submitted to INCRA by means of an official letter or by request from any interested party, and the self-definition will be entered in the General Register of the FCP, which is responsible for issuing the certificate. It is the responsibility of the Ministry of Culture, through the FCP, to assist and accompany the Ministry of Agrarian Development through INCRA in land regulation actions to guarantee the preservation of the cultural identity of the remnants of quilombo communities, as well as to subsidize the technical work when there is a challenge to the identification and recognition procedure. Remnants of quilombos are guaranteed participation in all phases of the administrative procedure, directly or through their representatives.

Once INCRA has completed the fieldwork of identification, delimitation and occupational and notary surveys, it must publish a notice twice in succession in the Official Gazette of the Union and in the Official Gazette of the federal unit where the area under study is located.

All the phases of the administrative procedure will be regulated by Normative Instruction 57/2009. It will lay down administrative procedures for the recognition of quilombo communities.

The first step is the self-identification of the community and its certification in the General Register of Remnants of Quilombo Communities by the FCP. It is INCRA's responsibility, through administrative procedures, to initiate the identification and delimitation process, which must be defined in meetings with the community and INCRA's interdisciplinary technical group.

The Technical Identification and Delimitation Report (RTID) must be based on objective elements, covering cartographic, land, agronomic, ecological, geographic, socioeconomic, historical, ethnographic and anthropological information, with relevant information on the community's general and specific data.

The RTID, after all the phases for recognizing the remaining quilombo community have been carried out, must be published twice consecutively in the Official Gazette of the Union. At the same time, they must be sent to the bodies and entities within 30 days to present their opinions on matters within their respective competencies.

3.3.3 Distribution of quilombos in Sergipe

Faced with the whole process of colonization and the usurpation of rights, and with the territorial formation imposed by virtue of a national project, in 1988, with the centenary of abolition, the black movement began to question segregation.

In Sergipe, according to a survey by the MDS (Ministry of Social Development), there are 46 communities: Lagoa dos Campinhos and Pontal dos Crioulos (Amparo do Sâo Francisco), Mocambinho (Aquidaba), Maloca (Aracaju), Massombro and Pontal da Barra (Barra dos Coqueiros), Brejâo dos Negros (Brejo Grande), Caraibas (Canhoba), Coqueiral, Pirangi and Terra Dura (Capela), Povoado Forte (Cumbe), Assungue, Canta Galo, Cassungue and Curuanha (Estância), Catuabo, Maria Preta and Manoel Bernades (Frei Paulo), Negro (Gararu), Bongue (Ilha das Flores), Desterro (Indiaroba), Patioba (Japaratuba), Ladeiras and Matamba (Japoatâ), Camo do Crioulo and Matembe (Lagarto), Mussuca and Quintale (Laranjeiras), Cambaza (Pacatuba), Alagamar and Marimbondo (Pirambu), Serra da Guia (Poço Redondo), Zumbi (Poço Verde), Mocambo (Porto da Folha), Santo AntonioCanafistula (Propriâ), Forras (Riachâo do Dantas), Quebra Chipre - povoado Bela Vista (Riachuelo), Bode and Botequim, Cajazeiras, Pedra d'Agua, Povoado Rua da palha and Taboa (Santa Luzia do Itanhy) and Quibonga (Sâo Cristóvào).

The criteria used by the MDS are: self-declaration and a survey carried out by other bodies working together in quilombola communities in relation to the number of communities provided by INCRA data. There are around fourteen communities with open recognition processes at INCRA, according to Table 1.

TABLE 1 - COMMUNITIES WITH OPEN PROCESSES IN THE STATE OF SERGIPE BY INCRA

ORDER NO.	COMMUNITIES	MUNICIPALITIES
01	Lagoa dos Campinhos	Amparo do Sâo Francisco
02	Maloca	Aracaju
03	Pontal da Barra	Barra dos Coqueiros
04	Brejâo dos Negros	Brejo Grande
05	Caraibas	Canhoba
06	Pirangy	Chapel
07	Strong	Cumbe
08	Catuabo	Friar Paul

09	Desterro	Indiaroba
10	Patioba	Japaratuba
11	Slopes	Japoatâ
12	Mussuca	Orange groves
13	Luzienses	Santa Luzia do Itanhy
14	Serra da Guia	Poço Redondo

Source: INCRA - table of open processes, 2012. Elaboration: Nivalda Batista Santos

TABLE 2 - REMAINING QUILOMBO COMMUNITIES CERTIFIED IN SERGIPE FROM 2004 TO 2012

Nª ORDER	MUNICIPALITY	COMMUNITIES
01	Amparo do Sao Francisco	Campinho Lagoon
02	Aracaju	Maloca
03	Barra dos Coqueiros	Pontal da Barra
04	Brejo Grande	Brejao dos Negros
05	Canhoba	Caraibas
06	Chapel	Canta Galo
07	Chapel	Pirangi Farm
08	Chapel	Terra Dura and Coqueiral
09	Cumbe	Povoado Forte
10	Estancia	Curuanha
11	Estancia	Porto D' Areia
12	Friar Paul	Catuabo
13	Flores Island	Bongue
14	Indiaroba	Desterro
15	Japaratuba	Patioba
16	Japoata	Slopes
17	Orange groves	Mussuca
18	Pirambu	Alagamar
19	Pirambu	Aningas
20	Poço Redondo	Serra da Guia
21	Green Well	Junco Lagoon
22	Porto da Folha	Mocambo
23	Proprià	San Antonio Canafistula

24	Riachuelo	Quebra Chifre (Bela Vista village)
25	Santa Luzia do Itanhy	Povoados Bode, Botequim, Cajazeiras, Pedra Furada, Rua da Palha and Taboa

Source: FCP, 2012

CHART 3 - REMAINING QUILOMBO COMMUNITIES WITH

OPEN PROCESSES FOR ISSUING CERTIFICATES IN SERGIPE

ORDER NO.	MUNICIPALITY	REMAINING QUILOMBO COMMUNITY
01	Aquidabâ	Mocambo
02	Canindé doSâo Francisco	Rua dos Negros
03	Friar Paul	Manoel Bernardes
04	Siriri	Castanhal

Source: FCP, 2013

Elaboration: Nivalda Batista Santos

TABLE 4 - KILOMB REMOTE CoMMUNITIES

IDENTIFIED IN sERGIpE

ORDER NO.	MUNICIPALITY	REMNANT COMMUNITY OF QUILOMBO
01	Amparo do sâo Francisco	Pontal do Crioulo

Source: FCp, 2013

Elaboration: Nivalda Batista santos

Most of these communities are located in rural areas, but have common characteristics such as a family workforce, accumulative capital and live off what they produce in their fields, from assistance programs and begin the process of recognition with the involvement of non-governmental organizations, the church and representatives of political parties, which include the discussion of Article 68 of the 1988 Federal Constitution. This is when the community began to mobilize in search of the right to access the land, as stated in the article, which establishes quilombola recognition. (Figure 3 - Map of the location of the municipalities and Figure 4 - Map of Quilombola Communities Not Titled)

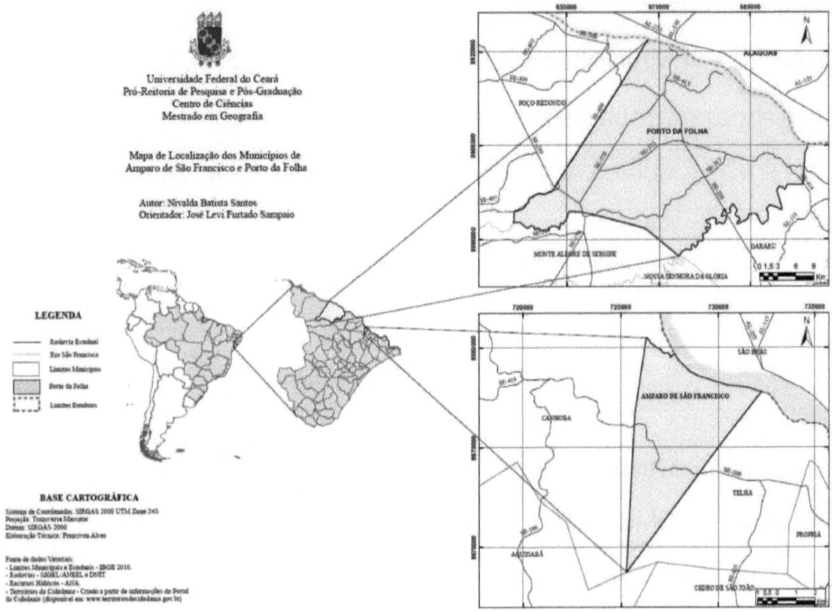

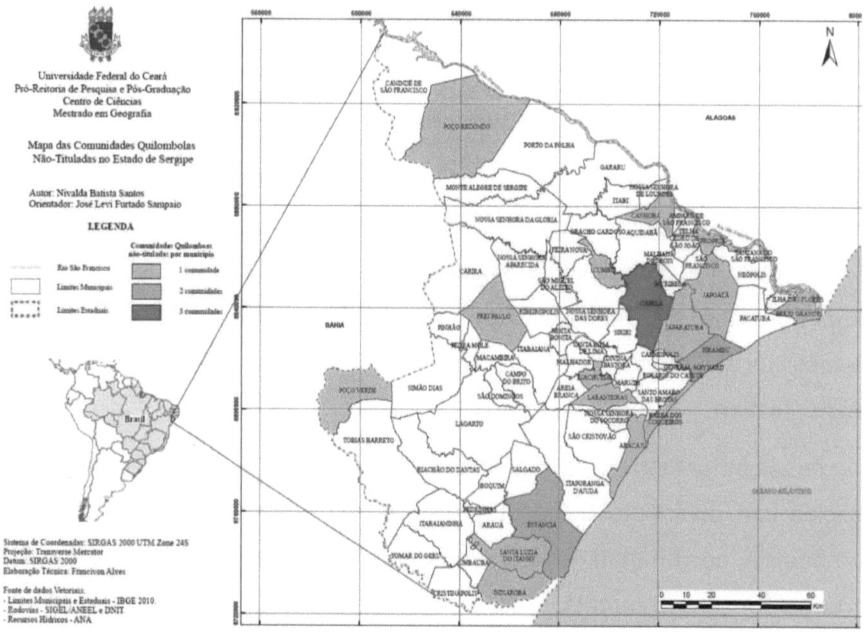

In today's scenario, to talk about quilombos is to reflect in the political and social context on issues that are still unclear, but which are moving in the direction of resolving and reducing the bureaucracy of the self-identification process, so that people can have access to the land.

As a reference point for a process of democracy in a country, state or nation, Chaui (2002) tells us that the ideas of equality and freedom as civil rights of citizens go far beyond their formal legal regulation. It means that citizens are subjects of rights and that where these rights do not exist or are not guaranteed, they have the right to fight for them and demand them. This is the core of democracy. But the contradiction between the social function of land for the quilombo remnants, the category under study, and for capital is reinforced by the state, which legitimizes landlordism, responsible for economic backwardness and the great exclusion of rural workers. This same state is indifferent to the power relations that take place in these territories of resistance and, linked to the judiciary, ignores the legitimate right of this group and defends the right to land ownership as an absolute and unlimited right of a minority, which is the ruling class.

3.4 Public Policies Aimed at Quilombo Remnants

Currently, there is a relationship of dependence on the public assistance policies used by public managers, which can either contribute or harm, since they prevent the population from seeking to fight for the rights they have acquired since the moment they exist, so they are unable to organize themselves and obey the orders of the representatives who impose and co-opt the population.

The development they preach is an increase in power and land income. Public policies can contribute when they are geared towards the interests of the community, the common good and adapted to regional conditions, rather than favoring dominant interests.

The culture of subalternity, in the form of clientelism and welfarism in force in the semi-arid region, hinders the struggle for public policies and local social organization in order to create the socio-political conditions for valuing and implementing an alternative model, through the demands of quilombo peasants. It is necessary to define a set of social and political practices in the life of the semi-arid population, such as cooperation, solidarity, mutual aid, etc.

Agriculture is the basis of sustenance for families living in rural areas, and provides the cultivation of products for the domestic market, as well as involving almost every family in working the land.

The state's primary function is to draw up and implement public policies that meet the objective needs and promote the well-being of society, and not to favor certain social groups and their interests.

The Agrarian Reform, based on the Land Statute created in 1964, has never been implemented, i.e., according to researchers, there has never been a nationwide Agrarian Reform aimed at changing the country's land policy. On the other hand, since the 1980s, the federal government has been implementing a policy of settling landless families benefiting from government programs that offer the necessary subsidies for the development of family farming due to the region's ecological conditions.

There is a high level of concentration of land and income, and a diagnosis of the land problems is necessary if there is to be any real implementation of agricultural policies. In the Northeast, the change in the distribution or structure of land ownership, which is generally the result of government initiatives, only aims to keep farmers on the land, without maximizing investment in credit. Depending on the social and political context, agrarian reform can have different purposes and take on different formats.

The agrarian reform that rural workers in general are demanding is a redistribution of income, power and rights, because workers want to see changes in the political and social structure of the country and the countryside, with farmers remaining in the place where they recognize themselves as people and can exercise their citizenship, abolishing all forms of exclusion.

These three items (public policies, poverty and basic needs) are interrelated in equal proportions, as there is a need to formulate public policies to eradicate poverty and offer the minimum conditions for survival in terms of basic needs.

Access to land and credit policies that stimulate production and guarantee income to overcome problems such as hunger, expansion of agricultural and land credit, renegotiation of small producers' debts, crop insurance, creation of a national system of technical assistance and rural extension, investments in infrastructure and support for marketing products. The use of mechanisms that guarantee a basis for sustaining family farmers, so that they are free from dependence on emergency and compensatory policies promoted by the government.

While regulatory policies seek terms in laws, decrees, regulations, ordinances that seek to regulate the behavior of individuals, companies and public entities, thus establishing the ways in which different social actors should relate; structuring policies, expressed in projects and actions that directly shape spaces and landscapes, and behavior-inducing policies, expressed in policies that seek to influence choices and behaviors of individuals.

The redesign of social protection systems began in the mid-1980s in some Southern Cone countries and at the turn of the 1990s in Brazil and other countries, pursuing common objectives: a) decentralization of programs and policies; b) greater community participation; c) targeting of the target public; d) greater concentration of social spending on investments in relation to costs; e) integration of programs and policies, amplifying their synergistic effects. In this focused welfare model, new forms of social protection are specified according to the type and degree of vulnerability of the target public, as opposed to universalist principles (FREITAS, 2007, p. 68).

Decentralizing power, fostering the formulation of public policies with society, opening the way for a critical reflection on agricultural relations, promoting a proposal based on valuing peasants who are fighting to remain in the countryside, and through the search for knowledge and the deepening of practices to improve the quality of life, making the most of the specificities and singularities of the place.

The Working Group of the Ministry of Social Development and Fight against Hunger (MDS), through the Multiannual Plan for the period 2004-2007, defines social assistance programs to serve the quilombola population. It puts the "organization of basic social protection in indigenous and quilombola communities" on the agenda of the 5th National Assistance Conference in 2005.

The Indigenous and Quilombola Communities WG sought to engage in dialogue with federal bodies such as the MDS, SNAS (National Secretariat for Social Assistance), government agencies, researchers from the black movement and advocacy organizations, and especially with national entities representing quilombola communities (Report of the Quilombola Population WG, 2007).

The actions carried out by the MDS in quilombola communities are divided into axes of action, which are: social assistance; conditional cash transfers - bolsa familia; actions in the field of food and nutritional security, combating hunger and malnutrition; and institutional coordination and partners.

From 2003 to 2006, Food Security distributed food baskets to quilombola communities, increasing the number of families benefiting each year in 19 states.

TABLE 5 - Distribution of Food Baskets

YEAR	BASKETS
2003	10.350
2004	35.432
2005-2006	260.312

Source: Quilombola Population WG Report, 2007

Under an agreement between the MDS/FCP and the Ministry of Culture (Minc), state actions were implemented in 150 remaining Quilombola communities, with R$5,000,000.00 in funds, aimed at helping approximately 15,000 families to become self-sustainable and overcome food insecurity. To this end, a socio-economic and cultural survey was carried out.

The conditional cash transfer represented by the registration of families from quilombola communities in the Bolsa Familia program.

Institutional coordination and partners that include various programs developed by other ministries, such as the Brazil Quilombola Program (whose coordinators are members of federal administration bodies and secretariats, such as the MDS, SEPPIR, MDA, FCP and INCRA), articulate government actions for the remaining quilombola communities and seek to draw up sustainable development plans for them.

Petrobras - Zero Hunger, projects approved and financed with R$3.8 million (Petrobras), with monitoring by the MDS and SEPPIR. The communities that benefited: Castanho (PE), Itamatatuia (MA), Ivaporanduva (SP), Mocambo (SE), Oriximinà (PA), Rio das Contas (BA), Sumidouro (PI), Fazenda Machadinha (RJ), Tapuio (PI), Campinho da Independência (RJ). And the MDS as managing member of the ethno-economic solidarity development project for quilombola communities (MDS TEM, MDA, SEPPIR, FCP, FUNASA/MS and FUBRA).

In the programs implemented by Social Assistance, there is no data to identify the quilombola community, but in 2004, 38 projects were approved by the MDS and agreements were signed to receive funds from the Comprehensive Family Care Program. In Sergipe, 5 municipalities received the PAIF to provide priority assistance to the remaining quilombo communities.

TABLE 6 - PAIF Resources for Quilombola Communities in Sergipe

MUNICIPALITY	UN	FAMILIES	RESOURCE/YEAR	AMOUNT/MONTH	VALUE/YEAR Jul/Dec
Aquidabâ	1	200	72.000	6.000	36.000
Chapel	1	200	72.000	6.000	36.000
Estancia	1	200	72.000	6.000	36.000
Gararu	1	200	72.000	6.000	36.000

Source: SNAS/MDS

These municipalities continued to receive these funds during 2005 and 2006. There has been an increase in the number of municipalities in Sergipe co-financed by PAIF, following the new rules defined by the National Policy of SUAS - the Unified Social Assistance System.

The Brazil Quilombola Program, created by SEPPIR in 2003, has a policy of guiding "quilombola communities, giving them greater objectivity in the search for legal, budgetary and operational obstacles". Its actions are carried out by 23 agencies of the federal public administration and are divided into four areas: land regularization; infrastructure and services; economic and social development; and social control and participation.

The main actions of the Brazil Quilombola Program cover the following dimensions:

a) Training of agents representing the remaining quilombo communities;

b) Promoting local development for quilombo communities;

c) Health care for quilombola populations, under the responsibility of the Ministry of Health;

d) Support for the sustainable development of quilombola communities, under the responsibility of the Ministry of Agrarian Development;

e) Payment of compensation to the occupants of lands demarcated and titled to the remnants of quilombos;

f) Recognition, demarcation and titling of areas

remaining quilombos, both of which are the responsibility of INCRA;

g) Support for the distribution of didactic and paradidactic material for primary schools and schools located in the remaining quilombo communities;

h) Support for training elementary school teachers to work in the remaining quilombo communities; (PBQ, 2008)

The effective consolidation of the actions implemented by the PBQ requires the involvement of the federal, state and municipal spheres and the participation of civil society. It was necessary to build a quilombola social agenda to articulate the actions carried out by the federal government, and its main areas of action are access to land, infrastructure and quality of life.

We can see the involvement of various federal bodies in implementing public policies that involve the remaining quilombo communities, but the resources invested are insufficient to serve the various communities that exist in Brazil. And in Sergipe, looking at the information available from the agencies and secretariats, public and social policies still have a welfare function.

Public policies need to be developed with the participation of the remaining quilombo communities, involving them in debates and discussions about what they need based on the difficulties they face in their territories. Building policies in the areas of health, education and the economy with them and for them. Policies that promote their inclusion, seeking strategies to reduce social inequalities and not compensatory policies.

4 MOCAMBO COMMUNITY FROM THE PERSPECTIVE OF ITS RESIDENTS

In this section we will link the theoretical discussions on the subject with the fieldwork carried out in the Mocambo Quilombo Community. With the rescue of the theme related to the issue of land ownership by traditional peoples, which still has some interpretations in the field of legal knowledge regarding its regularization in relation to article 68 of the ADCT and articles 215 and 216 of the Federal Constitution.

4.1 Characterization of the socio-territorial process

The municipality of Porto da Folha is located in the micro-region of the lower Sâo Francisco in the west of the state, with a territorial area of 895.1 km^2 . It is bordered to the north by the river Sâo Francisco and the state of Alagoas, to the south by the municipality of Monte Alegre, to the east by Gararu, and to the west by the municipality of Poço Redondo. It is the third largest municipality in the state of Sergipe, 190 km from the capital and situated at an altitude of 40 meters.

The relief is gently undulating, with the predominance of the Sertanejo pediplano, with flattened areas that gradually rise between 150 and 300 meters as you move towards the state of Bahia, and the presence of residual hills is quite common.

The predominant vegetation is hyperxerophilic caatinga, with an open, dense shrub formation adapted to the climate. There are other types such as capoeira, clean and dirty fields. Deforestation is occurring with the replacement of vegetation cover by pastures, favoring soil degradation and the emergence of areas with aspects of desertification.

The hydrography of the municipality of Porto da Folha is part of the São Francisco river basin, with rivers, streams and lagoons. It is made up of the Capivara, Campos, Cachorros and Mocambo rivers, as well as the Bebé, Comprida and Porteiras lagoons. Due to climatic conditions, some of these streams are temporary or intermittent. The municipality covers 33.5 km along the banks of the River San Francisco, with the following villages located there: Ilha do Ouro, Mocambo and Niterói.

As far as the climate is concerned, it is markedly semi-arid, with rainfall varying between 500 and 700 mm during the year, contributing to the cyclical period of drought between 7 and 8 months without rain, which contributes to the average annual temperature of 26°C.

Figura 5 - View of part of the community's territory during the dry season

Source: Field research, 2013

Figura 6 - View of the community area and the presence of the caatinga

Source: Field research, 2013

The municipality of Porto da Folha makes up the territory of the Alto Sertao Sergipano with six other municipalities: Canindé do Sao Francisco, Gararu, Monte Alegre de Sergipe, Nossa Senhora da Glória, Nossa Senhora de Lourdes and Poço Redondo. This territory has a population of approximately 137,926 inhabitants, with illiteracy rates of around 37%, a mortality rate of 23.6%, an HDI of 0.575 and a per capita income of R$ 75.00.

The predominant economic activity is agriculture based on permanent and temporary crops with the use of irrigation systems, which produces bananas, guavas and mangoes in the irrigated perimeter, located in the municipalities of Canindé do Sao Francisco, Poço Redondo and Porto da Folha. The most productive temporary crops are broad beans, beans and corn. Livestock includes large cattle and medium-sized pigs, goats and sheep.

The main commercial centers of the Alto Sertao territory are located in the municipalities of Canindé do Sâo Francisco and Nossa Senhora da Glória. In this territory there is a strong presence of agribusiness and the promotion of ecotourism due to the Xingó dam and the Sâo Francisco River Canyon.

This area has the highest concentration of settlements, with approximately 3,138 families settled in a space with little water availability due to environmental conditions, but there is a dominance of pastureland due to cattle ranching.

The municipality of Porto da Folha has a population of approximately 27,146 inhabitants. Its population density is around 30.94 inhabitants/km^2 , and its HDI is 0.556. Its economic activity is based on cattle breeding and the cultivation of corn, beans and manioc.

4.2 The Importance of Territory and Memory in the Process of Quilombola Identity

This dissertation argues that the recognition of the land ownership of the Quilombola Remnants goes through the symbolic character of the institutionalization of the territory by the community, through the relationships it develops from the moment the population appropriates and reproduces itself in that place.

In order to make this territory a reality and give legitimacy to their possession of the land they have conquered, it was necessary to contextualize the community's oral history, based on identity and memory, from the process of territorial and historical formation, and its self-recognition as a quilombo remnant. And, in this sense, to seek legitimization and land titling, as expressed in the article of the constitution.

According to Bosi (2003), memory is an instrument of mediation between our generation and the witnesses of the past. Through the memories of objects and images, we relate the concrete space of things to time.

The connection between sciences such as anthropology, sociology, history and geography is extremely important in order to contribute to the process of land regularization.

The territorialization of the remaining quilombo communities involves a new process of territorial planning based on the redistribution of land. Whether by expropriation or otherwise, it will characterize and bring out the contradictions of this territory and demonstrate the practices hidden behind the colonial territorial formation model adopted in Brazil.

The quilombola community is located in a semi-arid region where there are prolonged periods of drought due to the scarcity of water resources, despite being located on the banks of the São Francisco River, creating a relationship of dependence on public policies that are implemented in order to minimize the economic and social conditions in which they live.

In this process of construction and production of space, the quilombola peasant is trying to change his reality by seeking ways of living in the semi-arid region, even though ideology, through the discourse of difference, has imposed and placed the area as unattractive, inhospitable and with a needy population dependent on welfare policies.

The Community of Quilombo Remnants seeks, through its recognition, to transform and modify the historical and territorial formation imposed by those who had dominant power, through the recognition and titling of lands that guarantee their permanence in their territory, where economic, social, cultural and political relations are configured.

In today's context, in order to apply for quilombola identity, it is essential that there is a territory in which political, social, economic and cultural activities take place between the people who inhabit it, and that there is a need for a relationship of exchange. The quilombola identity of the Mocambo community was built on the effective action of the destruction of the rice paddy. For years, 40 families lived in the community alongside local farmers and some even worked on their properties.

According to Moraes, space represents the overlapping of natural results that coexist in contemporary times. In other words, it represents the community's conditions of existence and the territoriality that takes place with the relationships that are established from the logic of mobility, the information and knowledge that is sought in order to promote and rescue the historical aspect, in order to contribute to its insertion.

The notion of territory appears as a component that unites the action of individuals, at a time when territoriality expresses the preservation of identity, which represents a form of territorial ordering. In this way, by appropriating it, the individual or community conceives of it for themselves and for others.

After the destruction of the gardens, the community was unable to carry out their daily activities, as they were forbidden to do any kind of work in the area and were even prevented from crossing the borders of the territory, as they would be passing through the landowners' land, and they felt fear and hunger due to the lack of work. This prohibition was effectively enforced by the presence of jagunços hired by the landowners in order to secure possession of the land by force.

This action by the ranchers violates Article 5 (XV) of the Federal Constitution, which states that movement in national territory is free in times of peace, and that any person may, under the terms of the law, enter, remain in or leave it with their belongings (p. 6). The community was even prevented from moving around and communicating, since they couldn't use the access road.

4.2.1 Memory and history of the Quilombola Community of Mocambo

The importance of memory in the community is the narrative of those who have lived and kept in their memories the facts that represent their lives and experiences, which, with modernity, tend to be lost because we don't have the habit of listening to others. The narrator, for Walter Benjamin, is the one capable of narrating and telling his own story and that of others.

[...] The narrator is one of the masters and sages. He can draw on a lifetime of experience (a lifetime that doesn't just include his own experience of others. The narrator assimilates into his innermost substance what he knows from hearsay). His gift is to be able to tell his life [...] (BENJAMIN, 1994, p. 221).

According to Silva (2010), the resurgence of historical narrative and the memory of the forgotten necessarily involves listening to the past presented by the other. It means listening to the history of the other that isn't told in books and that has been forgotten by a society that doesn't care about the records of those who remained on the margins of the territorial formation process.

The remaining quilombola community seeks understanding in its own history of appropriation of the territory they claimed as their own and, through memory, seeks to rescue and invoke the past that has been passed down from generation to generation. As Benjamin (1987) says, articulating the past historically doesn't always mean knowing it as it really happened, but it does mean appropriating a reminiscence as it flashes up at the moment of danger.

The proof that the territory they now inhabit was previously occupied by Quilombos was "the lime stone foundations of non-existent houses and the remains of domestic utensils such as plates, glasses and bottles. It is these surface archaeological sites that the Mocambo people call pigsties, I.e. the places where their ancestors kept their old ranches" (ARRUTI, 2006, p. 264).

The way their ancestors used the land before occupation by the cattle ranch. The ranches represent the materiality of the appropriation of the place and the relationships that were established for subsistence by "free peasant farmers-fishermen" (ARRUTI, 2006, p. 265).

The community of Mocambo, located in the municipality of Porto da Folha, on the right bank of the river Sâo Francisco, has typical characteristics of riverside communities whose land they have occupied for generations. These lands were occupied from the 19th century onwards by blacks, who were known as "negros do pé de serra". Initially, they did not live on the banks of the Sâo Francisco, but were scattered throughout the area, due to the act of Emperor Pedro II who, when visiting Brazil in 1859, saw some houses on Ilha do Ouro and visited these communities living on the banks of the river, so he decided to demarcate their lands.

Approximately in 1882, after the Emperor's visit, an engineer was sent to demarcate the area that would be the indigenous mission and to demarcate all the lands of the Sâo Francisco, as his aim was to regularize or set up cattle ranches in the region. As a result of this demarcation, the farms Araticum, Mocambo, Jaciobà and Floresta came into being, which made up the dispute over possession and use of the territory between the remnants and the landowners.

One of these demarcated farms brought together in its area the spaces known as pigsties and the rice ponds occupied by the ancestors of the Mocambo community.

According to the memory of the community's elders, in the Quilombo territory there was a center known as "terra de heréu" where the ranches and borders were located. This was where the arable land was located due to its proximity to the river. From 1920 onwards there was a new delimitation of the land made by the state, and the appearance of fences caused by the demarcation of the farms that were located on the banks of the river contributed to the displacement of the area occupied by the squatters that was on the banks of the São Francisco River.

With the demarcation, the process of expropriation of the land occupied by the black families began. Their land was between the riverbank and what they planted was in sharecropping with the owners of the farms.

The possibility of the community keeping livestock was made impossible by the prohibition of the owners of the farms, who would only let them go if they paid for the lease of the land, thus contributing to the families moving away to more distant places.

The territory that expressed social relations was then squeezed between the farms, with the reduction of land, making both planting and animal husbandry unfeasible. In order to carry out either of these activities, they had to ask the landowner's permission, paying a rent or a stocking fee.

Rice was planted in the lagoons through the sharecropping system, in which the landowner gave the area and the seeds, while the work and other necessary expenses were borne by the squatter.

The children were responsible for keeping animals and birds away from the rice field. When the rice was harvested, it was spread out and divided in half, with the owner of the farm being responsible for cleaning it up, and what remained with the squatter was deducted from the value of the seeds that had been distributed. Everyone was responsible for their own activities, whether it was planting the rice, where the women made the beds, or harvesting, which was done by everyone.

In telling the story, according to Silva (2010: 68), "imagination emerges as an attempt to unveil and interpret, which can bring to light a past that, although it doesn't coincide with it, serves to clarify it". The narrative shows, through speech, how each individual participated in the work, whether in planting or raising animals.

In this relationship of antagonism between the squatters and the landowners, with the expansion of their land and the veiled imposition of their dominion, the squatters were forced to ask permission for any activity they were going to carry out in a territory that had previously been in common use, but had become, through the fences, limited.

It's worth noting that this limitation pushed the blacks to build Rua do Alto, even though one of the owners of the farm, who accepted the construction of the houses, prevented them from doing so.

The memory of playing songs when they were at the lagoon tending the rice field, songs that were passed down from generation to generation, as well as the coconut samba, a representation of the culture inherited from their ancestors. As described below:

Oh black man, let's get some coconut

Those who can't bear to take coconuts from the coconut tree

You can't stand fighting

Talk ox, sailor, sea

Talk ox, sailor, sea

Father of the family when he comes in from the fields

Run to the bakery

Buy pasta, bread, coffee

His son jumps into the yard I want money

To eat, says Dad on the carousel

My train of canes to make my victory

I'm taking a passenger freight train from viçosa

When you arrive in Vila Nova

On the whistle

My steam at the touch has a gallop

My steam has a whistle

When you leave for Paraguay

Speak ox, sailor of the sea (author unknown)

These songs were also sung when they gathered to commemorate the end of the mutirâo house-building process, at the time when the arruados of Alto and Mocambo emerged. They are expressions of the group's culture and territoriality.

4.2.2 The history of the quilombola community

According to the trajectory described by the deliberate conflict by the owner of the land used by the community, it was necessary for the remnants to organize themselves in order to decide how they would act after the destruction of their gardens.

As the CPT had already intervened in the municipality to advise rural workers in the struggle for land ownership, the Diocese of Proprià and the CPT tried to mobilize civil society organizations to discuss ways of resolving the conflict. This meeting was only attended by 80 members of the community, a church representative and a councillor and trade unionist (STR) from the municipality of Porto da Folha (CPT/Activity Report, 1992).

At this meeting, a committee of six community representatives was set up with the task of surveying the families' losses and, together with the priest and the councillor, going to the police station to file a complaint.

At this point, the process of organizing and politically training the peasants began, under the guidance of the CPT, in which the CPT proposed, through its representatives, that they be responsible for making decisions, based on the accumulation of decision-making power.

However, in this process of organization, fear set in because there were people in the community who continued to work for the landowners or people related to them, who represented the eyes and ears of the landowners and were considered spies, which is why it wasn't possible to hold the meetings in the community itself.

In order for these meetings to continue to take place, it was necessary to find another location, and the suggestion was indigenous territory (CPT/Activities Report, 1993).

These moments of mobilization served to prepare the community for the hearing with the judge in Porto da Folha over the constant arbitrary attitudes promoted by representatives of the police and the landowner's foremen, as well as the CPT's proposal to recover the community's history.

That's when the questions arose: Who are we? Are we sharecroppers? Peasants? Or quilombo descendants? On the basis of these concerns, the CPT presented the discussion for the community's knowledge of art. 68 ADCT of the 1988 Federal Constitution, which deals with quilombo communities.

At this meeting, some mobilization actions were proposed so that the peasants could decide how to deal with the following issues: 1. Remain in the disputed area, forcing the landowner to reach an agreement on labor compensation; 2. Demand recognition as "quilombo remnants"; 3. Evaluate the total area needed and request expropriation from INCRA, for the purpose of land reform.

The decision on which path to follow was only taken after a meeting with the Xocós, asking for their support (CPT/Activities Report, 1992).

This report also shows that the CPT ran a training course entitled "Young Leaders" in the Mocambo Quilombo. A "young leader" was appointed to represent the community and take part in actions related to the community, with the task of seeking support from institutions in Aracaju.

One of the courses of action decided upon jointly by Mocambo/Xocós was to occupy the disputed area, with the police intervening and the community mobilizing. In order to prevent the police from arriving, they picketed the road with stones (CPT/Activities Report, 1993).

The leadership training course provided participants with "explanations about agrarian reform, the nature of legal proceedings and their risks, as well as the formation of a Mixed Mocambo/Xocós Commission with local authorities, such as the District Judge, the Municipal Delegate, the Public Prosecutor, the INCRA State Prosecutor, the Corregidor and the Vice-President of the State Justice". (CPT/Activities Report, 1993).

In 1994, the CPT again organized a meeting asking for the support of civil society organizations in Aracaju, but this time it managed to bring together a good number of entities and decided to ask the History Department of the

Federal University of Sergipe (UFS) for support in carrying out a historical survey of quilombos in Sergipe.

Following this meeting, the black movement in Sergipe, represented by the SACI (Sociedade Afro Sergipana de Ensino e Cidadania) organization, created a project called "Movimento de Defesa do Mocambo" (Movement for the Defense of Mocambo) with the aim of overcoming fragmented actions. Within this project there was the prospect of creating a collective garden, but even though the money was available, it wasn't carried out as planned (Projeto Mocambo, 1994).

With the participation of representatives of the Mocambo community in regional meetings and the visit of the FCP, an association was created with a model statute structured especially for remnant communities.

Throughout the community's mobilization process, with the aim of resolving the land conflict in which they were involved, there were various retaliations on the part of the landowners, which were reported in the most important newspapers in the state of Sergipe, such as the headlines in the newspaper Gazeta de Sergipe for the 29th, in which it was announced on 30/05/1994: 1. **Jagunços ameaças colonos no interior**. The text refers to the tense atmosphere on Fazenda Sâo Francisco between its owner and the Mocambo families and the hiring of jagunços to terrorize the families. According to the newspaper report, the families lived as settlers. However, they were squatters because they had lived on the land for some time and were claiming their quilombola identity;

2. **Squatters in conflict with the courts in Porto da Folha.** The aim of this news is to distort reality, since the squatters were demanding the right to fair compliance with the Resolution. They were denouncing the arbitrary actions of the landowners to the courts.

In this article there is a statement from a Mocambo representative who says: "We want justice. The Judiciary has been slow, but we still believe in the Judiciary".

It was in this climate that the community remained in search of the regulation of their land, being threatened and prevented from exercising their right to come and go due to the presence of armed thugs. In the meantime, an injunction was issued by the Porto da Folha District Court for the evacuation, which was immediately revoked. However, on May 17, 1994, another eviction injunction was issued, but the squatters' legal advisors filed a request for repossession.

Looking at the newspaper articles and the documents that reveal the history of the struggle, we can see the distortion of the media's objectives in relation to the squatters' struggle. The media defend the dominant power. There was every interest in hiding the real situation of the squatters' struggle and the threats suffered by their leaders. It camouflaged the real intention of the landowners to be able to appropriate the lands of the Quilombo remnants, through the manipulations of local and state political leaders, with the aim of subjugating the Mocambo.

The main objective was to present the population with a critical situation of conflict, of tensions, in an attempt to curb constitutional rights. It is important to emphasize that at that time the community was going through a serious situation, because, in addition to the lack of conditions for survival due to the semi-arid climate in the period of the drought cycle, there was also the lack of work, since they were prevented from cultivating the land. The only way they could survive was on pensions from families with elderly relatives.

In opposition to this situation, the CPT and the Sergipe Public Prosecutor's Office asked the FCP to carry out a historical survey of the quilombo in order to guarantee the legitimization and recognition of the remaining quilombo community.

The Attorney General's Office has been involved in the struggle of the black community of Mocambo, and has been involved in clarifying and enabling access to land for the remnants in accordance with Article 68 of the ADCT.

This even required the creation of the Antonio do Alto association and the request for recognition through an application to the FCP by a member representing the community.

In 1995, the FCP, together with the Brazilian Anthropology Association, sent an anthropologist to Mocambo to carry out a survey and produce an anthropological report on the community with the aim of officially recognizing the group as quilombo remnants.

"The report began in 1997, through agreement No. 132/96 - SE of the Center for Studies on Territory and Traditional Populations and the Ministry of Culture, and was coordinated by the FCP and carried out by Eliana Catarina O" Dwyer - PhD in Anthropology; José Mauricio Andion Arruti - anthropologist; and Francisco Lopes de Souza - Agronomist. (DOU, 28/05/97) using ethnography to explain the occupation of the territory and the development of the activities carried out by the squatters and their living relationships.

On May 28, 1997, the FCP approved the Report of Identification of the Remaining Community of Quilombo (DOU, 28/05/97), consisting of the plan and descriptive memorial. But the title of recognition of ownership was only granted to the remnants of the Mocambo community on July 14, 2000 by the FCP.

Mocambo is the first Quilombo remnant community in Sergipe to be recognized after almost a decade of struggle to obtain land titles with the participation of organizations, providing legal advice and guidance on how to deal with each demand that arose. In addition to the moments of tension and threats they suffered so that they would back down from their goal of obtaining ownership of the land and exercising their condition of being peasants.

After recognition, some projects were developed in the community through the availability of resources being developed by non-governmental organizations (NGOs), such as Caritas Diocesana de Poço Redondo, which carried out projects with resources from the Dom Helder Project, funding from the Federal

Government and PETROBRAS.

The projects developed through the Antonio Alto association involve raising cattle, goats, chickens and a vegetable garden.

At the moment, some families are running their own farm, where in the late afternoon those responsible put on water and feed, and the vegetable garden project where they grow coriander, manioc and yellow and purple elephant grass.

Basic sanitation is non-existent in the community, and drainage occurs above ground and towards the banks of the river. Garbage is incinerated by the community. There is no sewage system or garbage collection, which contributes to the emergence of diseases.

All the houses have piped water and only occasionally do they use the river for washing dishes and bathing - this happens when there is a shortage of water.

As for health, the families said that there is no medical care in the community, although there is a health worker to attend to them, he generally doesn't make home visits and doesn't receive the materials needed to monitor the families. The quilombo community seeks care in the municipalities of Nossa Senhora da Gloria/SE and Pâo-de-Açùcar in the state of Alagoas and, in emergencies, in hospitals in Aracaju,

Although the municipality has Community Health Agent Programs and a Family Health Program, the doctor usually doesn't come to the community. The neglect of municipal public health is palpable. The health center located in the community is completely abandoned and has not been properly maintained by public managers.

Figura 7 - Health center in a state of disrepair due to lack of maintenance by public managers

Source: Field research, 2013

Figura 8 - Health center

Source: Field research, 2013

As far as education is concerned, there is a Quilombola State College in the community, which offers primary and secondary education to students from quilombos and nearby communities such as: the Paulo Freire Settlement, the

Vitoria do Sâo Francisco Settlement, Bela Aurora, Esperança, Fazenda Santa Fé, Fazenda Sâo Joâo, Fazenda Sâo José, Jureminha and Pedro Leâo, which account for 50% of the students enrolled from these localities.

The school's staff is made up of 17 members: 10 teachers (10% from Aracaju, 10% from Mocambo, 10% from Proprià and 70% from Pâo-de-Açùcar), 3 cooks (67% from Pâo-de-Açùcar and 33% from Mocambo), 2 security guards (Pâo-de-Açùcar) and 2 services (Mocambo) provided by the town hall.

The presence of people from the city of Pâo-de-Açùcar in Alagoas, which borders the Mocambo Quilombo, can be seen.

In 2013, 184 students were enrolled at the school, divided into primary (104) and secondary (80) grades.

CHART 7 - NUMBER OF STUDENTS PER GRADE IN elementary SCHOOL

SERIES	NUMBER OF STUDENTS	%
1ST YEAR	07	6,7
2ND YEAR	10	9,6
3RD YEAR	09	8,7
4TH YEAR	02	1,9
5TH YEAR	10	9,6
6TH YEAR	25	24,0
YEAR 7	08	7,7
7ª SERIES	14	13,5
8ª SERIES	19	18,3
TOTAL	104	100

Source: Field research, 2013.

According to data from the field research, there is a low level of education among adults and the elderly, although they have the knowledge acquired through experience and involvement in actions involving issues related to the community.

In relation to the school's infrastructure, there is a lack of attention to the maintenance of the building, inadequate classrooms for students, damaged toilets and drinking fountains, a digital inclusion laboratory that is not working, despite having the equipment, and a lack of an internet network that makes it impossible to use.

The school is housed in two buildings with seven classrooms to cater for all the daytime and afternoon students, one of which combines the 4th and 5th grade classes due to the number of students enrolled.

During the school term there are a number of projects that are part of the pedagogical project being carried out at the school: Human values such as ethics, respect, solidarity and justice; the black consciousness project, which focuses on the appreciation of black people and the history of Zumbi, with the participation of the community through coconut samba and capoeira.

Figura 9 - Project panel on human values developed at the school

CHART 8 - NUMBER OF STUDENTS PER GRADE IN SECONDARY EDUCATION

SERIES	NUMBER OF STUDENTS	%
1ST YEAR	30	37,5
2ND YEAR	30	37,5

5TH YEAR	20	25
TOTAL	80	100

Source: Field research, 2013

Although the external structure of one of the buildings looks good, it is internally that the problems are very visible, denoting the real conditions of the space in which the dissemination of knowledge is taking place.

Figura 10 - Side view of the community school with the satellite dish

Source: Field research, 2013

Figure 11 - View of the school's inclusion laboratory

Source: Field research, 2013

Figure 12 - View of the access door to the school's inclusion laboratory

Source: Field research, 2013

Figure 13 - Quilombola State College - building 2

Source: Field research, 2013

Frequent leisure activities in the community are: going to mass, chatting in front of the houses at night, playing soccer in the late afternoon, and bathing in the river, especially at weekends when the family gets together.

Figure 14 - Football in the late afternoon

Source: Field research, 2013

The feast of the Holy Cross, held every May, lasts 10 days, with nine novenas. Each night, groups are honored: cowboys, fishermen, charcoal burners, children and young people from the community. The opening ceremony on the first day begins at 5am with a workshop and at 8am there is a performance by a coconut samba group made up of members of the community. The festival encompasses the community and the surrounding areas.

Figure 15 - Front view of the church

Source: Field research, 2013

Figure 16 - Paved path to the church

Source: Field research, 2013

The community was assisted with social programs such as the construction of cisterns through the One Million Rural Cisterns Program (P1MC), as well as the construction of a paved cistern to collect rainwater for human consumption and planting. It also benefited from the construction of 97 houses, 80 in Mocambo and 17 in Ranchinho.

Figure 17 - Cisterna Calçadâo in the community of Ranchinho

Source: Field research, 2013

A cistern built by the One Land, Two Waters Program (P1+2), implemented by the ASA (Articulation of the Brazilian Semi-Arid), its purpose is to capture rainwater to be used for planting and animal consumption.

The association has some of the equipment used by the community, which came from the resources of projects financed by PRONESE, the Dom Helder Câmara Project and Petrobras, such as a tractor and a motorcycle, as well as the renovation of the clubhouse, the vegetable garden and the farm. The regularization of documentation for access to PRONAF (National Program for Strengthening Family Farming) and the retirement of 20 quilombola descendants, as well as the distribution every two months of food baskets by the FCP to all registered families.

With the resources invested in the territory, it was possible to buy machinery for a flour mill, build a garage and a field with the goalposts. In addition to the refurbishment of the association's administrative headquarters and the purchase of computers.

After 20 years of fighting for the regularization of their land, the Mocambo

Quilombo Community is stagnating, with no prospects of immediate change. Despite their inclusion in social programs, they are currently unable to carry out projects due to the association's administrative problems.

The Quilombo is experiencing yet another cycle of drought, as has already been observed, due to its location in a region with a semi-arid climate. There is an inhospitable environment in the territory, resulting in the loss of the basic crops of corn and beans. The quilombolas are surviving on grants from social programs, such as the family grant offered by the federal government. The inactivity of young people and adults is contributing to the rise in alcoholism.

The majority of public positions in the education sector are held by people from other municipalities, in this case mostly from Pâo-de-Açùcar in Alagoas, who return at the end of the working day due to the proximity. It is only necessary to cross the river Sâo Francisco.

The community also seeks other services such as health care, higher education and commerce (Figure 18).

Figure 18- Bots on the banks of the River San Francisco

Source: Field research, 2013

The Mocambo community is often linked to the municipality of Pao-de-Açúcar in Alagoas in search of health and education services and facilities, in addition to the existing kinship relationship that causes this displacement.

FINAL CONSIDERATIONS

The fight for access to land in Brazil by quilombola communities aims to ensure their permanence in the territory and also raises a much-discussed issue due to the concentration of the land structure based on the colonization model implemented in the country.

In the 1980s, the Unified Black Movement emerged in Brazil, raising serious questions about the social injustices disseminated in Brazilian society against black people, stemming from the colonization process implemented in the country with the use of black people as a productive force in sugar cane plantations in the Northeast and coffee plantations in the Southeast.

It's worth pointing out that Brazilian society is very fragile when it comes to building a collective conscience, conquering and establishing respect for others. This is why it has historically established ethnic boundaries, and why it is necessary to fight to reduce the social injustices that have persisted for centuries.

Even with the recognition of quilombola communities in Brazil and in Sergipe, despite the inclusion of art.68 ADCT of the 1988 Federal Constitution, it took a decade for the right of possession to be regulated, and actions by the Black Movement were necessary for the National Constitution to prescribe it.

In order to prove their quilombola identity, blacks need to recover through memory the relationships that their ancestors had with the territory, establishing their territoriality at different historical moments.

The memory of the search for the past will provide instruments to meet the criteria that characterize the request for land ownership, through the identity of the remnants of quilombos, providing information in the sense of reconstructing the territory used by their ancestors through the symbolic references of the bonds of coexistence, and the social, economic and political relationships developed in the movement back and forth through the feeling of belonging.

The quilombola issue, after more than 20 years of Article 68, which governs the right to land ownership in the territories of black communities, is still a slow-moving issue. From the moment the protagonists self-identify and the recognition process is opened, it takes years for the remnants to actually obtain land ownership. They are exposed to the tensions imposed by those who have, in their favor, ways of preventing the black community from obtaining its own autonomy in its family production unit. The land is the extension of their lives, the means of meeting their needs as individuals and as peasants.

The Mocambo community has been assisted by infrastructure programs such as paving the road to the church, building affordable housing, installing a drinking water network and buying materials and equipment to carry out agricultural activities.

The community's struggle for access to land has helped the representatives, based on their knowledge of what it means to be a quilombola, to take part in local, regional and national discussions, contributing to the group's political formation and strengthening in the quest for recognition.

The destruction of the rice paddy became the milestone for the community, after decades of being subjected to the owners of the farms, to mobilize to take back their territory in order to maintain their practices and customs.

The struggle for land regularization in Mocambo has helped other communities in the state to claim recognition as quilombolas, demanding the institutionalization of their territories so that, through self-identification, they can gain access to and possession of the land.

The construction of the movement in defense of Mocambo has provided visibility for issues related to access to land for black communities, with actions by the state's Black Movement through links between institutions.

After being recognized as quilombola communities, the social policies developed to assist quilombola communities, in order to ensure that instruments can be incorporated to enable their autonomy, are presented as compensatory policies. The programs implemented by the different state and federal public administration bodies do not always have common objectives, after the process of titling the lands of the remaining quilombolas.

The goals and resources of each program fall short of meeting the needs of the quilombola communities in the states, especially those located in semi-arid climates that have water shortages, making it impossible to grow crops or use them for human consumption.

These social policies minimize the living conditions and permanence of the community.

The struggle for the land of labor (use value) against the land of business under the domination of agribusiness cannot advance in isolation, individually and locally. The challenge is to understand that it is only possible through the collective struggle of rural and urban subjects. The state, through compensatory public policies, before guaranteeing the right to land ownership, seeks to find forms of mediation to avoid confrontations with the big land and capital owners and to guarantee forms of articulation with the market for capitalist accumulation.

REFERENCES

ABA. Documents of the Working Group on Rural Black Communities.**Boletim Informativo NUER**, n. 1, 1994.

ANJOS, Rafael Sanzio Araùjo. **Spatial distribution of the remaining quilombo communities in Brazil.** Humanities Magazine. Brasilia: Editora UNB, n° 47 nov. 1999.

ANTONIL, André Joao - **Cultura e opulência do Brasil por suas drogas e minas** - Sao Paulo - Ed. Nacional - 1967.

ARRUTI, José Mauricio A. The emergence of "remanescentes": notes for dialogue between indigenous people and quilombolas. **Mana**, Rio de Janeiro, n. 3-2, 1997.

ARRUTI, José Mauricio.**Mocambo:** anthropology and history of the quilombola formation process**.** Bauru, SP: Edusc, 2006 (Social Sciences Collection).

AZEVÊDO, Eliane. **Race, concept and prejudice**. Sao Paulo: ATICA, 1987 - Spring Series.

BADIE, B. O fim dos territórios: ensaio sobre a desordem internacional e sobre a utilidade social do respeito. Lisbon: Instituto Piaget, 1996.

BEZERRA, Feltre. **Etnias Sergipanas: contribuiçâo ao seu estudo**. 10ª Ed. Aracaju: Gràfica Editora J. Andrade, 1984

BOSI, Ecléa. ***Memory and society* - memories of old people**. 3ed. Sao Paulo: Cia das Letras, 1994. 484p.

BOSI, Ecléa. **OTempo vivo da memória: ensaios e psicologia social, Sâo Paulo**: Ateliê editorial, 2003.

BENJAMIN, Walter. **The narrator**. Considerations on the work of Nikolai Leskov, in Selected Works: Magic and Technique, Art and Politics: Essays on Literature and Cultural History, translated by Sérgio Paulo Rouanet, Sao Paulo: Brasiliense, 1994.

BRAZIL. Presidency of the Republic. Decree No. 4.887, of November 20, 2003. Brasilia - DF.

Decree No. 96999 of October 20, 1988. Available at: http://www.mma.gov.br/sitio/index. php. Accessed on September 29, 2011.

. Constitution (1988). Transitional Constitutional Provisions Act, article 68.

Report on the Identification of the Mocambo Quilombo Community. Official Gazette of the Union. 28/05/1997.

. QUILOMBOLA BRAZIL PROGRAM. Brasilia, SEPPIR 2008

Report of the Quilombola Population WG. Brasilia, MDS, March/2007.

CARNEIRO, Edson. **O quilombo dos Palmares**. Sâo Paulo: Ed. Nacional, 1988

CARNEIRO, Sueli. Gender and race. In: BRUSCHINI, Cristina; UNBEHAUM, Sandra G. (Org.).**Gênero, democracia e sociedade brasileira.** Sâo Paulo: Fundaçâo Carlos Chagas / Ed. 34, 2002. p. 167-194.

CASTELLS, Manuel. **The Power of Identity**. Translation by Klauss Brandini Gerhardt - Sâo Paulo: Paz e Terra, 1999. - (The information age: economy, society and culture; v. 2) 530 p.CPT. Activity Report. 21/02/1992.

CHAUi, Marilena. Conformism and Resistance: aspects of popular culture in Brazil. Sâo Paulo: Brasiliense, 1987.

CHAUi, Marilena. Presentation, in BOSI, Ecléa. ***Memòria e sociedade* - lembranças de velhos**. 3ed. Sâo Paulo: Cia das Letras, 1994, p. 17 to 33.

Portuguese Overseas Council of 1740.

CONCEIÇÀO, Alexandrina. Classroom notes for the course, History of Geographical Thought, given at NPGEO/UFS, 2004.

. **The Peasant Question: Looking at it from a Dialectical Point of View**. Master's dissertation. NPGEO/UFS, 1991.

FAIRCLOUGH, N. **Discurso e mudança social**. Coord. translation, revision and pref. to the Brazilian edition by Izabel Magalhâes. Brasilia: Editora Universidade de Brasilia, 2008.

. **Analysingdiscourse**: textual analysis for social research. London; New York: Routledge, 2003.

FERNANDES, Florestan. **Integraçâo do negro na sociedade de classes**. Sâo Paulo: USP, 1964.

FIGUEIREDO, Ariovaldo. **O negro e a violência do branco: o negro em Sergipe**. Rio de Janeiro: José Alvaro Editora, 1977.

FIORIN, J. L. **Linguagem e Ideologia.** 8ª ed. São Paulo: Atica, 2007.

FREIRE, Felisberto - **História de Sergipe.** 2nd Ed. Petrópolis - Vozes - 1977.

FREITAS, Décio. **Palmares: the war of the slaves**. Rio de Janeiro: Graal, 1990.

Décio. **Escravismo Brasileiro**. 2nd Ed. Porto Alegre: Mercado Aberto, 1982.

. **The Quilombos and the Black Rebellion**. São Paulo: Ed. Brasiliense, 1981

FREITAS. Rosana C. Martinelli Freitas. **The Lula government and social protection in Brazil: challenges and prospects.** Ver. Katal. Florianopolis v. 10, p. 65-74 jan./jun. 2007.

HOORNAERT, Eduardo. The mass of the quilombos came too late?.**Tempo e Presença**, n.173, 1982.

IANNI, Octavio. **The metamorphoses of the slave**. São Paulo: Hucitec, 1988 (207271).

MARK, Karl. **Para a Critica da Economía Politica**, transl. José Arthur Gisnotti and Edgar Malagodi, in Pensadores, 2nd Ed. São Paulo: Abril Cultural.1983

MARTINS, José de Souza. **O Poder do Atraso: ensaios de Sociologia da História Lenta**. São Paulo: Hucitec, 1994.

MARTINS, José de Souza.**A chegada do estranho**. Sao Paulo: Hucitec, 1993.

MARTINS, José de Souza. **A Caminhada No Chao da Noite**: Emancipaçao politica e libertaçao nos movimentos sociais no campo. Sao Paulo: Hucitec, 1989.

MARTINS, José de Souza. **Os Camponeses e a Politica no Brasil.** Petrópolis: Vozes, 1983.

MINAYO, Maria Cecilia de Souza (org.). **Pesquisa Social**: teoria, mètodo e criatividade. Petrópolis: Vozes, 21ª ed, 2002.

MORAES, Antonio Carlos Robert. **Geografia Historica do Brasil:** cinco ensaios, uma proposta e uma critica, ia Ed. Sao Paulo: Annablume, 2009

MORAES, Antonio Carlos Robert. **Territory and History in Brazil**. 2nd Ed. Sao

Paulo: Annablume, 2005.

MORAES, Antonio Carlos Robert. **Geografia Pequena História Critica**. 2nd Ed. Sao Paulo: Hucitec, 1983.

. **Introduction**. In: Ratzel. Sao Paulo: Atica, 1990.

MATTOSO, Kàtia de Queiroz. **Being a Slave in Brazil**. Sao Paulo: Ed. Brasiliense, 1982.

MORAES, Antonio Carlos Robert; COSTA, Wanderley Messias da.**Geografia critica - A valorizaçao do Espaço**. Sao Paulo: Hucitec, 1984.

MOREIRA, Rui. **Da regiao à rede ao lugar (a nova realidade e o novo olhar geogràfico sobre o mundo**). In: Ciência Geogràfica, ano III, no 6. Bauru: AGB, 1996.

MOURA, Clóvis. **Rebellions from the slave quarters: quilombos, insurrections, guerrillas**.

Sao Paulo: Ed. Conquista 1972.

Clovis. **Quilombos: resistance to slavery**. Sao Paulo: Atica, 1987

MOURA, Margarida Maria. **Peasants**. 2nd Ed. Sao Paulo: Atica, 1988.

NASCIMENTO, Abdias. **Quilombismo**. Petrópolis: Vozes, 1980

NUNES, Maria Tetis. **Colonial Sergipe I**. 2ª Ed. São Cristôvâo: Editora UFS; Aracaju: Oviêdo Teixeira, 2006

INCRA and the Challenges of Regularizing Quilombola Territories - Some Experiences. Brasilia, MDA/NEAD, 2006

INCRA. Journal of Agrarian Law and Land Reform. - Year 20 n° 20 - Brasilia: INCRA, 2007

OLIVEIRA, Ariovaldo Umbelino de. **Recent Territorial Transformations in the Brazilian Countryside**. Examination for the position of Full Professor. FFLCH=USP. 03/03/1998.

, Ariovaldo Umbelino de. **Peasant Agriculture in Brazil**. Sâo Paulo: Contexto, 2001.

RAFFESTIN, Claude. **For a Geography of Power**. São Paulo: Atica, 1978.

SANTOS, Milton. **Territory: Globalization and Fragmentation**. 3rd Ed. Sâo Paulo: Hucitec, 1996.

RESENDE, Viviane de Mello and RAMALHO, Viviane. ***Critical Discourse Analysis.*** Rio de Janeiro: Contexto Publishing House, 2009.

ISTO É MAGAZINE. Sâo Paulo, p. 34, June 20, 1990.

. **The Nature of Space, Technique and Time, Reason and Emotion**. 2nd Ed. Sâo Paulo: Hucitec, 1997.

UNDP BRAZIL. *Atlas of Human Development in Brazil, 2000.* Available at: http://www.pnud.org.br/atlas

SACI/SE. Mocambo Defense Movement Project. 02/02/1994

SANTOS, Milton. Territory: Globalization and Fragmentation. 3rd Ed. Sâo Paulo: Hucitec, 1996.

. **For a New Geography**. Sao Paulo. Hucitec, 1978.

SERGIPE. State Government. *Digital atlas of Sergipe's water resources;* Sergipe's water resources information system - SRHSE. Aracaju, n.d., CD-ROM.

. State Government. Sergipe Territory Development Plan. 2008. 40p.

. State Government. *Sergipe in Data 2008*. State Planning Secretariat. 110 p.

. State Government. *A state of big opportunities*. [s.n.t.]

. State Government. FEDERAL UNIVERSITY OF SERGIPE. *State of Sergipe: A proposal for territorial planning*. Aracaju, 2007. V. 1 and 2.

. Development plan for the Alto Sertao Sergipano territory. 2008. 88 p.

SILVA, Maria Ester Ferreira da - **A Desterritorializaçao do Povo Xucuru- Kariri e o Processo de Demarcaçâo** de **Terras Indigenas no Municipio** de **Palmeiras dos Ìndios,** Master's dissertation. NPGEO/UFS, 2004

, **Territory, Power and the Multiple Territorialities of Indigenous and Black Lands: narrative and memory as a means of constructing the territory of traditional peoples**. Doctoral thesis. NPGEO/UFS, 2010.

SILVA, Sylvio Bandeira de Melo; SILVA. Barbara Christine Nentwing. **Studies on Globalization, Territory of Bahia.** UFBA. Master's Degree in Geography, Department of Geography, 2003, p.182.

SMITH, Roberto. **Land ownership**. Sao Paulo: Brasiliense, 1990

SOUZA, Marcelo Lopes de. **Territory over space and power, autonomy and development**. In: CASTRO, Inà Elias de. (org.) Geography: Concepts and themes. Rio de Janeiro: Bertrand Brasil, 1995.

SOUZA, Maria Adélia Aparecida de. **Geographies of Inequality: Globalization and Fragmentation.** In: SANTOS, Milton. Territory: Globalization and Fragmentation. 3rd Ed. Sâo Paulo: Hucitec, 1996, p. 21-28

WEBER, Max. **Economy and Society: Foundations of Comprehensive Sociology.** Translated by Regis Barbosa and Karen Elsabe Barbosa. Brasilia: Editora Universidade de Brasilia, 1999.

APPENDIX

APPENDIX A - DATA COLLECTION INSTRUMENT

FIELD RESEARCH QUESTIONNAIRE

Date: // ___LOCATION: ____

Name of interviewee:____

Age: years Sex: ( ) Male ( ) Female

Are you from this town? ( ) Yes ( ) No.

If NOT, where did you live? ____

Number of family members?

How many children? ___________

Are you a family farmer? ( ) Yes ( ) No

You are: ( ) a squatter ( ) a sharecropper ( ) a family farmer ( ) a tenant ( ) a salaried worker ( ) other ____

Do you live solely from farming? ( ) Yes ( ) No Why?

____

How many people work? __________________

What is your total income? ( ) less than 1 salary ( ) 1 to 2 salaries ( ) 2 to 3 salaries ( ) 3 to 4 salaries ( ) more than 4 salaries

Is the property? ( ) owned ( ) loaned ( ) occupied

( ) rented

How big is the property? ____

What types of crops?____

Do you keep animals? ( ) Yes ( ) No Which ones? _________________________

Do you belong to? ( ) association ( ) union ( ) other organization. which one?_____

____

What leisure activities are there in the community?

Do you receive any benefits from a social program ( ) Yes ( ) No

Which ones?

With the recognition of the quilombola community, what changes have you seen?

The community has public and private projects ( ) Yes ( ) No Which ones?

Which institution is responsible for the execution?

What results have been achieved?

Such as the public services provided by municipalities, states and the Union:

Education

Health

Water supply

Sanitation

basic

What kind of activities are young people involved in in the community?

What prospects do you have for the community after titling and possession?

GENERAL COMMENTS

APPENDIX B - DATA COLLECTION INSTRUMENT

RESEARCH SCRIPT

RECOVERING MEMORY

Date: // ___LOCATION: ____

Name of interviewee:____

Age: years Sex: ( ) Male ( ) Female

What was the community like before recognition?

What image do you have of the community?

What activities did you do before? And how?

What does it mean to be a remnant of a quilombo? Quilombola?

What are some of the highlights of this community that you could mention?

What changes do you think are positive or negative for the community as a result of the recognition?

What is the Mocambo community like today?

MIX
Papier aus verantwortungsvollen Quellen
Paper from responsible sources
FSC® C105338

Printed by Books on Demand GmbH, Norderstedt / Germany